Gauss's Law

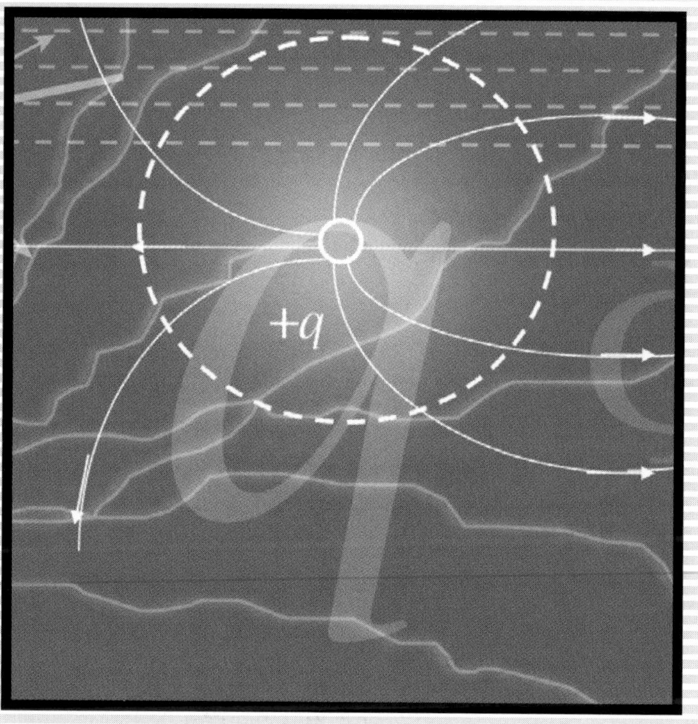

by Marion Appelquist

For my husband Tom

Publisher: Tom Maksym
Executive Editor: Steven Jay Griffel
Editor: Richard deFurla
Copy Editor: Dee Josephson
Vice President of Production and Manufacturing: Doreen Smith
Production Director: Nicole Dawson
Production Editor: Hugh Haggerty
Art Director: Eric Dawson

PEOPLES® PUBLISHING GROUP

ISBN 1-4138-2208-8

299 Market Street
Saddle Brook, New Jersey 07663

Printed in the United States of America.

10 9 8 7 6 5 4 3 2 1

TABLE OF CONTENTS

Introduction

Karl Friedrich Gauss (1777–1855) has been called one of history's most brilliant mathematicians, certainly on the same level as such giants as Archimedes, Riemann, and Sir Isaac Newton. Among Gauss's many contributions to pure mathematics are his work on complex numbers and non-Euclidean geometry. Like Newton, Riemann, and Archimedes, Gauss also worked on problems that we would now call physics. He developed a theory of optics and did calculations of the orbits of planets, an area known as celestial mechanics. Gauss knew of Newton's work and sought to develop for electromagnetism a theory as simple and elegant as Newton's theory of gravitation. He did not succeed fully, but his work stands as an essential part of the foundation of the classical theory of electromagnetism developed by James Clerk Maxwell in the 1860s. Maxwell's great synthesis of the work of Gauss, Ampere, and Faraday would not have been possible without the work of Gauss. The focus of this book is the equation known as Gauss's Law for Electric Fields, often shortened to just Gauss's Law.

In Chapter 1, a brief review of charge and electric fields is presented. This is not meant to be a comprehensive treatment of electric fields, as many of the familiar integrations involving rings and lines of charge are not included. These methods are touched on only briefly through the calculation of the field due to a sheet of charge, which allows a contrast to the use of Gauss's Law to solve the same problem. Still, you must know how to do such integrations, because methods using Gauss's Law are not appropriate for every problem. They are most applicable to problems having certain types of symmetry. Chapter 1 moves on quickly to a mostly descriptive justification of Gauss's Law, leading to an equation you will see used throughout the rest of the book. The equation relates the charge enclosed within an imaginary surface to the electric field at all points on the same surface.

In Chapter 2, some general properties of Gauss's Law are explored along with some important applications to conductors. You will also see many examples in which Gauss's Law is used to determine the electric field for symmetric charge distributions by a much simpler method than standard integration. However, this law is more than just a simplifying tool. Understanding Gauss's Law will deepen your understanding of electric fields and how they behave. For example, why is it that excess electric charge on a conductor is always located on the outside surface? Using Gauss's Law, you will see how this can be explained simply and elegantly.

Chapter 3 begins with a brief review of electric potential energy, electric potential, and potential difference. Much use is made of the gravitational analogy because it is often easier to picture. This link is possible because of the mathematical similarity between Newton's Law of Gravity and Coulomb's Law for the force between two charges. Electric field and electric potential go hand in hand, not only on the AP test, but in a thorough understanding of electrostatics. Many of the charge distributions from Chapters 1 and 2 are revisited in Chapter 3.

Following each chapter, there are exercises and problems for you to do. Many of these are very similar to questions and problems that are part of the AP curriculum. The exercises generally require a simple answer of a verbal nature or a diagram. The problems require mathematical calculations involving vectors, algebraic manipulation, and calculus. Answers to the odd-numbered exercises and problems are found at the end of the book; answers to all exercises and solutions may be found on the publisher's Web site at http://www.peoplescollegeprep.com/gauss_law. These solutions give you the chance to check your work and to see if your understanding of electrostatics has been improved by reading this book. I hope that you will come to find this area of physics as fascinating as I do.

Marion Appelquist
New Haven, Connecticut

CHAPTER 1

Electric Fields and Gauss's Law

1.1 Review of Electric Charge

Electric charge comes in two varieties, positive and negative. Like charges repel each other and unlike charges attract each other. Electrons bear the negative variety of charge, and it is these particles that move from object to object when charging takes place. Rub a glass rod with silk, and electrons move from the glass to the silk. The glass is then said to be **positively charged,** meaning that it has more protons than electrons. The silk is **negatively charged,** having more electrons than protons. Or, rub a piece of hard rubber with wool; electrons will flow from the wool to the piece of rubber, positively charging the wool and negatively charging the rubber. In the lab you may have used charging strips made of acetate or vinyl, which acquire a positive or negative charge, respectively, when rubbed with almost anything, including paper towels.

Note that charge is neither created nor destroyed in such a process, but only redistributed. All the electrons leaving one substance go to another substance, according to the principle of **charge conservation.**

Imagine that you touch a positively charged strip to a small metal ball. The strip shares some of its charge with the ball, and then both are positively charged. This is an example of charging by contact. It is also possible to charge an object by the process of **induction**. Suppose, as illustrated in Figure 1.1, a positively charged strip is brought near a metal ball. Electrons will be attracted to the side of the sphere nearest the positive charging strip.

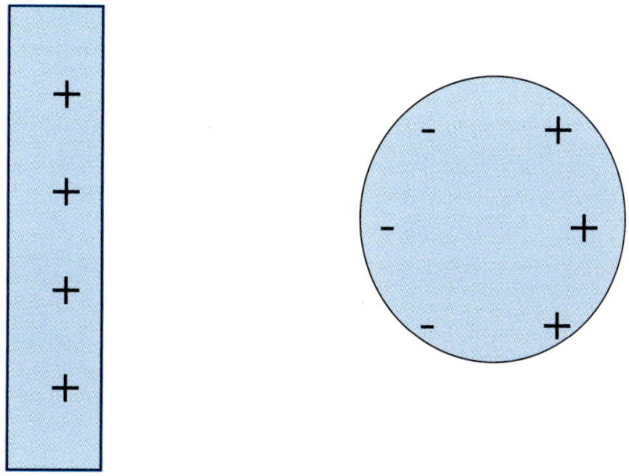

Figure 1.1

Now if you touch the metal ball with your finger, as shown in Figure 1.2, electrons will flow from your finger to the ball.

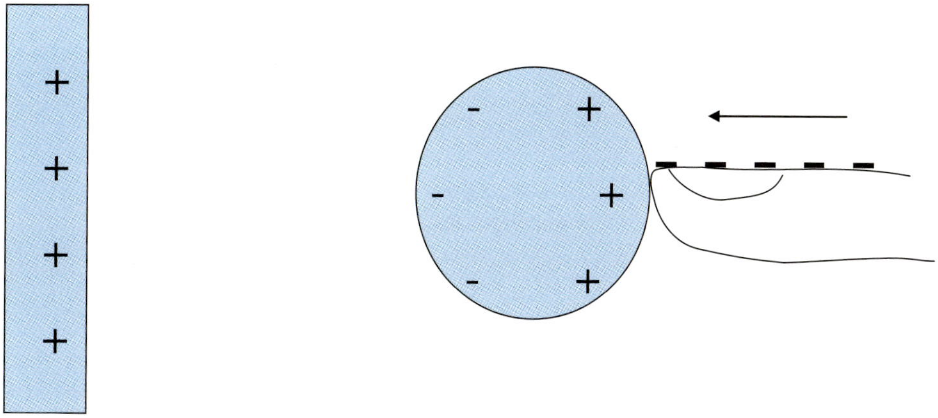

Figure 1.2

Remove your finger and then remove the strip. Now you have a conducting sphere with negative charge spread uniformly all over it, as in Figure 1.3. Although the charging strip never touched the ball, yet it was able to induce a charge on the ball.

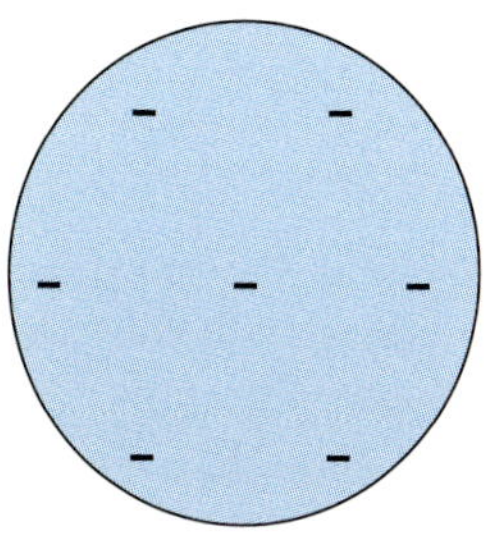

Figure 1.3

A **conductor** has many electrons that are loosely bound to their atoms and can be made to move around freely in the material. An **insulator**, by contrast, has its electrons tightly bound to its atoms or molecules; in these materials electrons cannot move freely. Figure 1.4 shows a piece of insulating material, sometimes called a **dielectric**, between two oppositely charged conducting plates. The dielectric reacts to the presence of the charges in a special way. Its molecules become stretched so that their sides nearer the positive plate bear a negative charge, and vice versa. There are, of course, many more molecules than suggested by the diagram. This device is called a **parallel plate capacitor**.

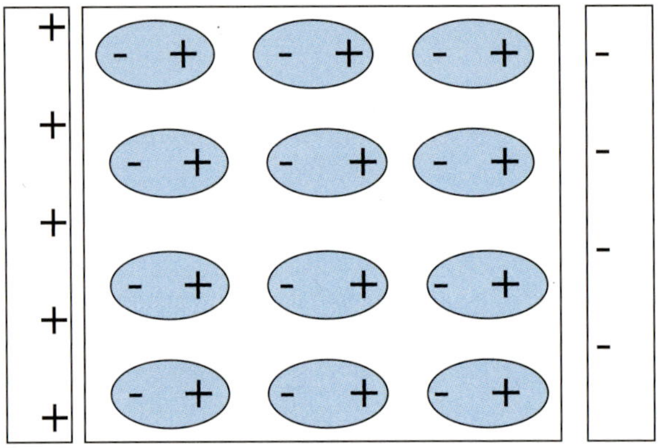

Figure 1.4
Conducting plates with dielectric.

The process of touching the metal ball in Figure 1.2 was an example of **grounding**. The same effect is produced by attaching a wire from the conducting ball to the Earth, so that Earth is then the source of the electrons instead of your body. A small charged conductor, whether touched by you or connected to Earth, will share its charge. Since both you and Earth are much larger, the small conductor loses essentially all of its charge to become **neutral** (or uncharged).

1.2 Coulomb's Law and the Electric Field

When two charges q_1 and q_2 are at rest, the magnitude of the force each of them exerts on the other is directly proportional to the product of the charges and inversely proportional to the square of the distance between them. This relation is called **Coulomb's Law** and in equation form becomes

$$F_{12} = \frac{kq_1q_2}{r^2}$$

The symbols q_1 and q_2 represent the magnitudes of the charges, and r is the distance between the charges. When standard units are used, charge is in Coulombs (C) and distance is in meters (m). The force is then in Newtons (N). The constant k has the value

$$k = 9.0 \times 10^9 \text{ N} \cdot \text{m}^2/\text{C}^2$$

in a vacuum. The direction of the force is along the line connecting the two charges.

There is another way to write the value of the constant k and, consequently, Coulomb's Law. It is sometimes convenient to redefine the constant k as

$$k = \frac{1}{4\pi\varepsilon_0}$$

where ε_0 is called the **permittivity of free space** (or the vacuum permittivity). The constant ε_0 is universal, just as k is. Its value is

$$\varepsilon_0 = 8.85 \times 10^{-12}\, \mathrm{C}^2/\mathrm{N} \cdot \mathrm{m}^2$$

We can then write Coulomb's Law as

$$F_{12} = \frac{1}{4\pi\varepsilon_0}\frac{q_1 q_2}{r^2}$$

The pair of forces that the two charges exert on each other must be equal and opposite, in accordance with Newton's Third Law. It does not matter which charge is the larger of the two, as the force is proportional to their product. Like charges repel and unlike charges attract, as stated in Section 1.1.

Supposing a single charge is placed in a vacuum. You might imagine, as the English physicist Michael Faraday first imagined, that an electric field reaching out to infinity is produced around the charge. The field may be represented by a set of lines, as shown in Figure 1.5. This line diagram represents a field with spherical symmetry, going out in all directions in space. By convention, the field is depicted as pointing in the direction of the force on one small, positive test charge placed in the region.

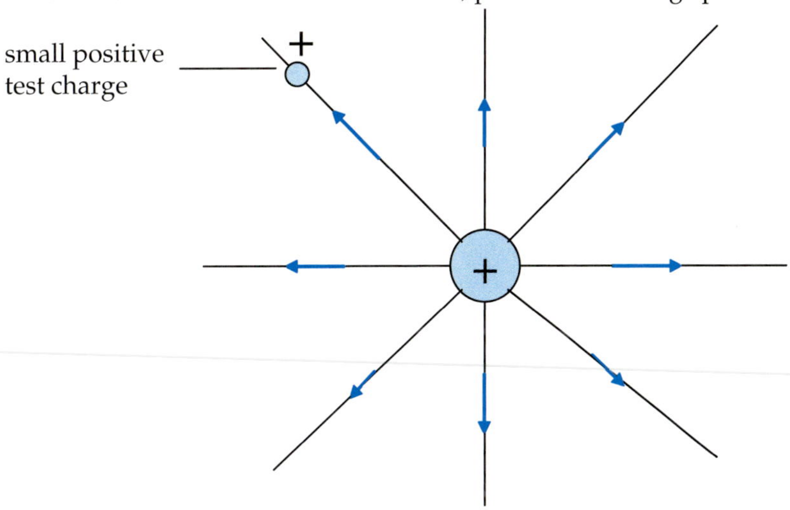

Figure 1.5

A field diagram for a single negative charge in a vacuum is shown in Figure 1.6. The only difference between this field and Figure 1.5 is that the arrows point inward.

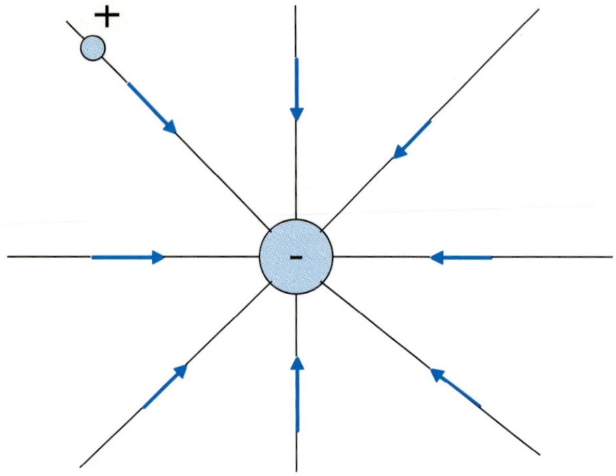

Figure 1.6

Note that a field diagram also indicates the relative strength of the field. Where the lines are close together—that is, relatively nearer the charge—the field is stronger. As the lines spread farther apart, the field becomes weaker.

Electric field strength is a vector quantity. The magnitude $\boldsymbol{E}$ of the electric field strength due to a point charge q is given by

$$E = \frac{1}{4\pi\varepsilon_0}\frac{q}{r^2}$$

The units of E are Newtons/Coulomb, abbreviated N/C.

A pair of equal and opposite charges would produce the field shown in Figure 1.7. Notice that the electric field lines always point away from the positive charge and toward the negative charge.

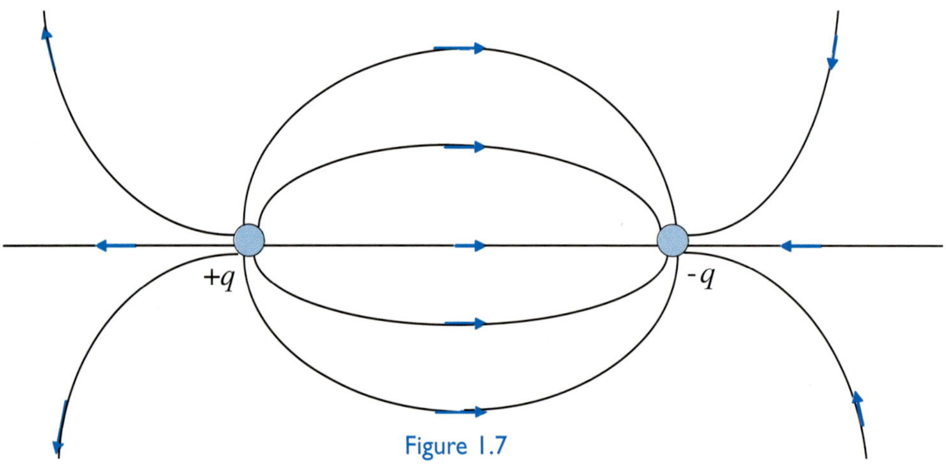

Figure 1.7

Unequal but opposite charges would produce a field looking like Figure 1.8.

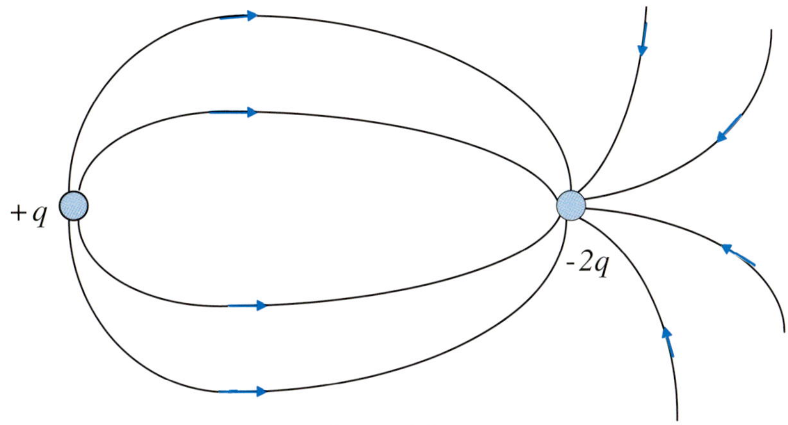

Figure 1.8

Note that the lines are more dense around the charge of greater magnitude. In fact, twice as many lines are shown entering $-2q$ as leaving $+q$.

Another interesting field configuration is produced by the device of parallel plates mentioned in Section 1.1. As shown in Figure 1.9, the field is uniform in the middle area between the plates, only deviating near the edges of the plates.

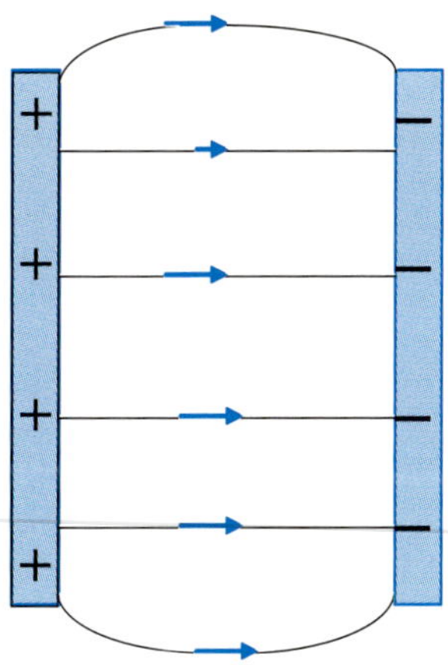

Figure 1.9

To find the magnitude and direction of an electric field at any point in space, for fields such as those shown in Figures 1.7 or 1.8, requires the addition of the vector fields due to each of the individual point charges. The direction of the resultant field will always be tangent to the field line at that point. Let's look at a specific situation.

Example 1 A charge $+q$ is located at the origin and another charge $-q$ is located at the coordinates (5, 0). The magnitude of q is 8.0×10^{-9} C. Find the electric field strength at a point P in space with the coordinates (2, 2).

Solution First, draw a good diagram, showing the approximate positions of the vectors. $\vec{E}_1$ is the vector contribution to the field from charge $+q$ alone. $\vec{E}_2$ is the vector contribution to the field from $-q$ alone. Electric field vectors obey the **superposition principle**: if you have more than one charge contributing to an electric field, the combined field is the vector sum of the fields generated by the individual charges. In this case, $\vec{E} = \vec{E}_1 + \vec{E}_2$, as illustrated in Figure 1.10. It is important to note that $\vec{E}_1$ and $\vec{E}_2$ are not perpendicular in this situation.

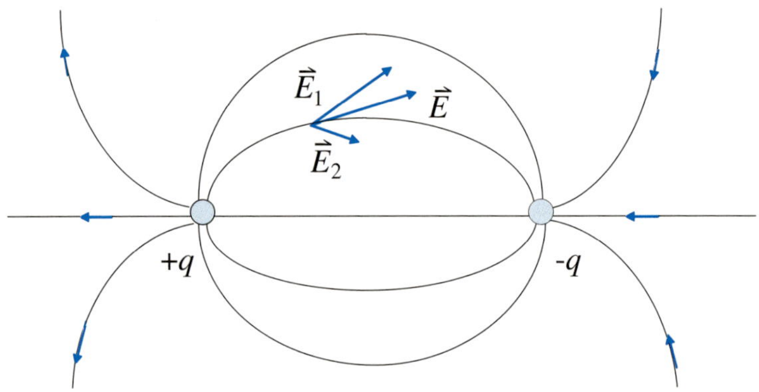

Figure 1.10

Next, you need a diagram that shows the distances and labels the angles in the problem (Figure 1.11).

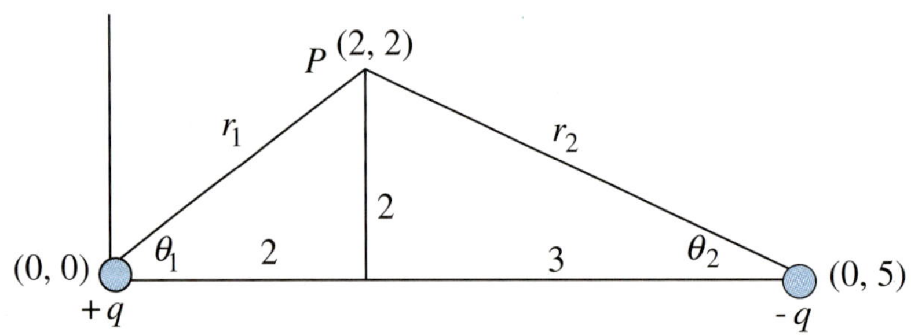

Figure 1.11

As we established earlier, the magnitude of $\vec{E}_1$ is given by

$$E_1 = \frac{1}{4\pi\varepsilon_0}\frac{q}{r^2}$$

$$E_1 = \frac{(9.0 \times 10^9)(8.0 \times 10^{-9})}{(2^2 + 2^2)}$$

$$E_1 = 9.0 \text{ N/C}$$

The vector $\vec{E}_1$ may be written as

$$\vec{E}_1 = 9.0(\hat{i}\cos\theta_1 + \hat{j}\sin\theta_1)$$

The angle θ_1 is 45°, so

$$\vec{E}_1 = 9.0\left(\hat{i}\frac{\sqrt{2}}{2} + \hat{j}\frac{\sqrt{2}}{2}\right)$$
$$\vec{E}_1 = (6.4\hat{i} + 6.4\hat{j})\text{N/C}$$

To find $\vec{E}_2$ you must first find the radial distance from the charge $-q$ to point P, and the angle this radius makes with the x-axis. Referring to Fig. 1.11, you can use the right-hand right triangle, together with the Pythagorean theorem, to get

$$r_2^2 = 3^2 + 2^2 = 13$$
$$\tan\theta_2 = \frac{2}{3}$$
$$\theta_2 = 33.7°$$

The magnitude of $\vec{E}_2$ is

$$E_2 = (9 \times 10^9)(8 \times 10^{-9})/13 \text{ N/C}$$
$$\vec{E}_2 = \frac{72}{13}(\hat{i}\cos 33.7° - \hat{j}\sin 33.7°)$$
$$\vec{E}_2 = (4.6\hat{i} - 3.1\hat{j}) \text{ N/C}$$

Finally, you add $\vec{E}_1$ and $\vec{E}_2$ to get

$$\vec{E} = (11.0\hat{i} + 3.3\hat{j}) \text{ N/C}$$

which seems to agree visually with the vectors illustrated in Figure 1.10. If this result were given in polar coordinates, the magnitude of $\vec{E}$ would be

$$E = \sqrt{11.0^2 + 3.3^2} = 11.5 \text{ N/C}$$

and its direction would be

$$\theta = \tan^{-1}\left(\frac{3.3}{11.0}\right) = 17°$$

Let's do one more example. You will find the results of it very useful for comparison purposes when we get to Gauss's Law itself.

Example 2 Imagine a large plate made of non-conducting material with positive charge spread uniformly all over it, like a piece of bread spread evenly with peanut butter. Find an expression for the electric field strength at a point P a distance y above the plate. Assume that the ratio of y to the plate dimension is small, but not infinitesimally small.

Solution First, imagine one ring of charge dq. (Later, we will be able to sum up all such rings.) The radius of the ring is x and its width is dx, as given in Figure 1.12.

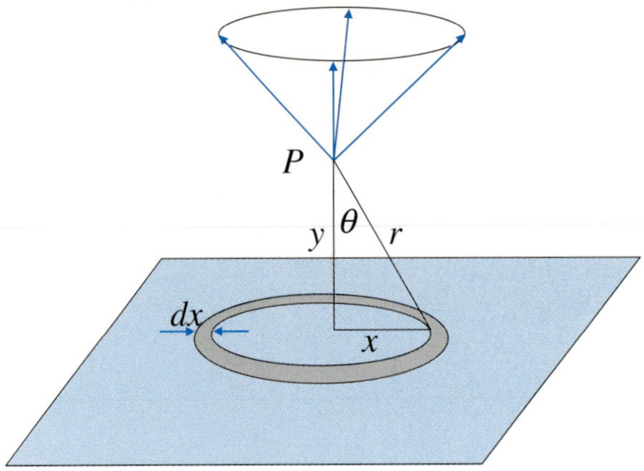

Figure 1.12

Each little portion of the ring produces a vector field at point P. You will notice that all the vectors from the ring form a cone. The horizontal components cancel out, leaving only the vertical components to be summed. We will call this sum dE.

$$dE = \frac{1}{4\pi\varepsilon_0}\frac{dq}{r^2}\cos\theta$$

The ring charge dq can be written as σdA where σ is the charge per unit area on the plate and dA is the area of the ring. To find dA, imagine cutting the ring and opening it up into a long rectangle. Its length is the circumference $2\pi x$ of the ring and its width is the width of the ring dx. Thus, the area is length times width.

$$dA = 2\pi x dx$$

Then, substituting σdA for dq

$$dE = \frac{\sigma x \cos\theta \, dx}{2\varepsilon_0 r^2}$$

You will notice that we have three different variables in this equation—x, r and θ. (Remember that y is a constant for a given choice of point P.) You will need to choose one of them and express the others in terms of the chosen variable. Let's choose θ.

$$x = y\tan\theta$$

$$dx = y\sec^2\theta d\theta$$

and

$$r = \frac{y}{\cos\theta}$$

Substituting, you get

$$dE = \frac{\sigma y^2 \tan\theta \cos\theta \sec^2\theta \cos^2\theta d\theta}{2\varepsilon_0 y^2}$$

$$dE = \frac{\sigma \sin\theta d\theta}{2\varepsilon_0}$$

Now, when the small positive charge placed at point P is affected by the ring of charge right below, the angle θ is zero. When it is affected by a ring of charge very far away on the plate, the angle approaches $\frac{\pi}{2}$. So we integrate over that interval.

$$E=\int_0^{\pi/2} \frac{\sigma \sin\theta d\theta}{2\varepsilon_0}$$

$$E = \frac{\sigma}{2\varepsilon_0}\left[-\cos\theta\right]_0^{\frac{\pi}{2}}$$

$$E = \frac{\sigma}{2\varepsilon_0}$$

This result is rather remarkable because it is a constant independent of y. It says that as long as you do not move point P too close to the edge of the plate, or too far above the plate, the value of E will be constant. We will return to this problem later and you will see how to derive the same result much more easily, using Gauss's Law.

1.3 Gauss's Law

Imagine a small light source in empty space. Photons of light emanate from the source and spread out evenly in all directions, as illustrated in Figure 1.13.

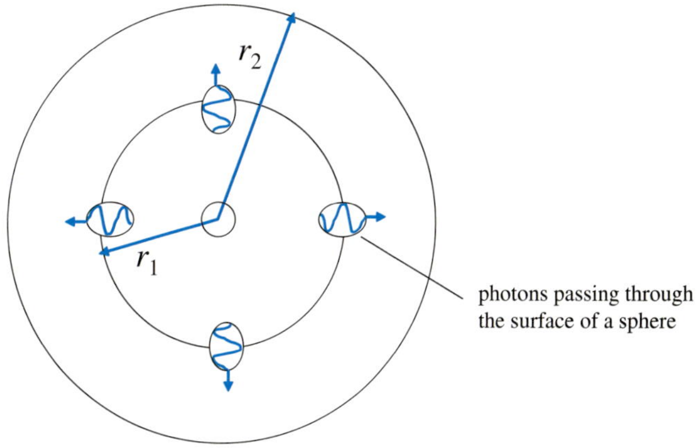

Figure 1.13

Assume that the source radiates photons at a constant rate, and imagine a sphere of radius r_1 that encloses the source. Every second the same number of photons will pass through the surface of this sphere. Now imagine a second sphere of radius r_2. As long as there is no absorption of photons, the same number will pass through the surface of this sphere every second as pass through the first sphere; however, the numbers of photons per second per square meter at the two radii will differ. In other words, the brightness or intensity will be greater at r_1 than it is at r_2.

The same conclusion may be reached by reasoning in this way. The surface area of a sphere is given by $A = 4\pi r^2$. Since the area increases as r^2, the number of photons per second per square meter must be inversely proportional to r^2; that is,

$$\frac{\#\text{ photons}}{\text{s}\cdot\text{m}^2} \propto \frac{1}{r^2}$$

When you multiply this value by the area, you will get a constant, the number of photons per second:

$$\frac{\#\text{ photons}}{\text{s}\cdot\text{m}^2}\cdot\text{m}^2 = \frac{\#\text{ photons}}{\text{s}}$$

Because photons per second is constant, the intensity of the light varies inversely as the square of the distance from the source. The truth of this relationship can also be verified by experiment.

But what if the surface is not as regular as a sphere? For example, imagine an irregular closed surface, like a potato skin with the potato scooped out. The surface might look something like Figure 1.14.

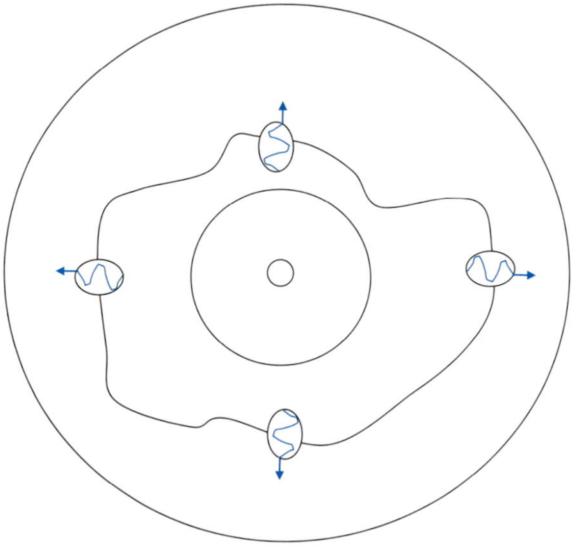

Figure 1.14

Once again, as long as there is no absorption, as many photons as go through the inner sphere each second will also go through the potato skin surface and, eventually, through the outer sphere as well. ***But note***: we have to be a little careful in talking about the intensity of the light *at a particular place* on an irregular surface, because the light is not necessarily hitting the surface at a right angle, as it is on the spheres.

To understand this distinction, let's look at a slightly different example of the behavior of light, where it strikes the surface as a **parallel beam**. (You may have talked about parallel beams in your study of lenses, or recall that the sun is so far away from Earth that its rays behave very much like a parallel beam as they strike the Earth.) Say that a certain number of photons per second is represented by the five lines shown in the next illustration. In Figure 1.15a the lines strike a small strip of area at a right angle. In Figure 1.15b, the same lines strike a strip of area obliquely; therefore, the area they strike is larger, and the light is not as bright on this second surface. (This is part of the reason that sunlight is so much more intense in the tropics than at higher latitudes.)

The dotted shape in Figure 1.15b shows a **projected area**, which is perpendicular to the rays and, in fact, the same area as in Figure 1.15a. This means that, if you use the projected area instead of the oblique area, the photons per second per square meter will be the same as in Figure 1.15a. By trigonometry, the projected area equals the actual area multiplied by the cosine of the angle θ, where θ is the angle between the two area strips.

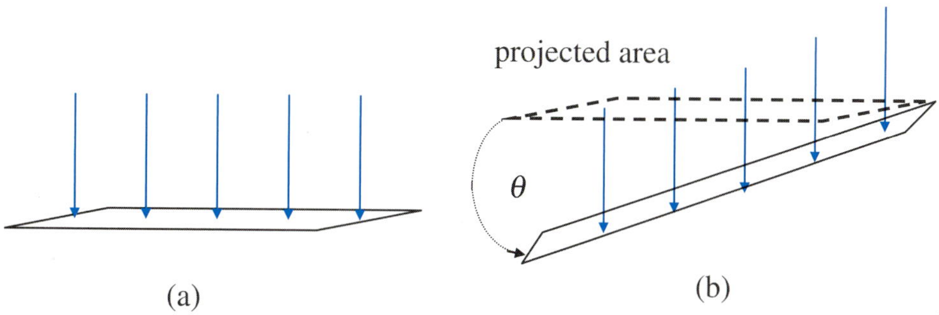

Figures 1.15a and 1.15b

Now let's go back to the situation in Figure 1.14 and look at it again—first, qualitatively. The light may be less intense on the parts of the irregular shape that are not perpendicular to the light, but this lower intensity is offset by the fact that the irregular shape has a larger surface area than a sphere with the same average radius. You can see why if, instead of thinking about the oblique areas, you focus on their corresponding projections, which are perpendicular to the rays of light.

The situation also needs to be looked at again quantitatively. Because the light emanates from a point source, the group of radial lines will first go through the small patch of area $\Delta\vec{A}_{1i}$ on the sphere of radius r_1, as shown in Figure 1.16, and then strike the patch of area $\Delta\vec{A}_i$ on the potato skin. (The meaning of the vector symbols on these areas will be defined shortly.) The projected area for $\Delta\vec{A}_i$ will be the area that is intercepted by these same radial lines, such that the radius $\vec{r}$ is perpendicular to the area. The projected area is actually part of a sphere of radius $\vec{r}$. Each little piece of projected area increases as r^2, while the number of photons per second per m^2 passing through it decreases as $\frac{1}{r^2}$. Thus we have arrived at the same conclusion for the potato skin as for the sphere. In fact, the number of photons per second will be constant for *any closed surface* that contains the light source within, no matter what its shape.

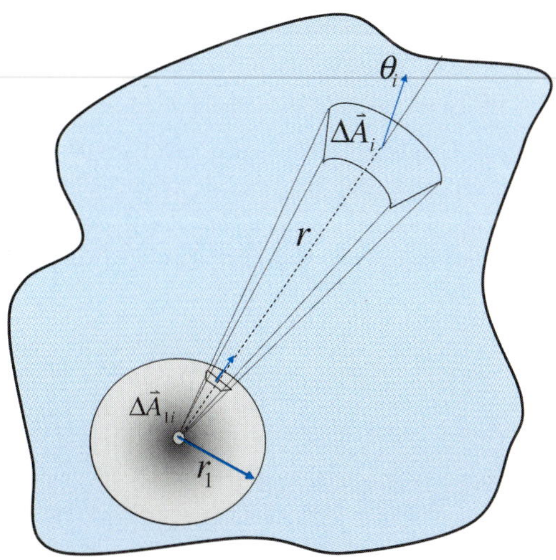

Figure 1.16

At this point we can make an analogy to the electric field. Like the number of photons per second per m^2, the electric field due to a point charge in empty space is inversely proportional to r^2.

$$E = \frac{1}{4\pi\varepsilon_0}\frac{q}{r^2}$$

Imagine surrounding the point charge by a sphere. To visualize, use Figure 1.16, but replace the photon source by a positive charge. The electric field points radially outward and the field vectors "cut through" the surface area of the sphere just as the photons passed through the sphere. How much of a vector field "cuts through" an area is measured by a quantity called the flux of the vector field—in this case, the **flux of the electric field**. To define the flux precisely, we must introduce an area vector. The area vector is defined such that it points perpendicularly to the area and has a magnitude equal to the area. In Figure 1.17, a uniform field cuts through a circular area.

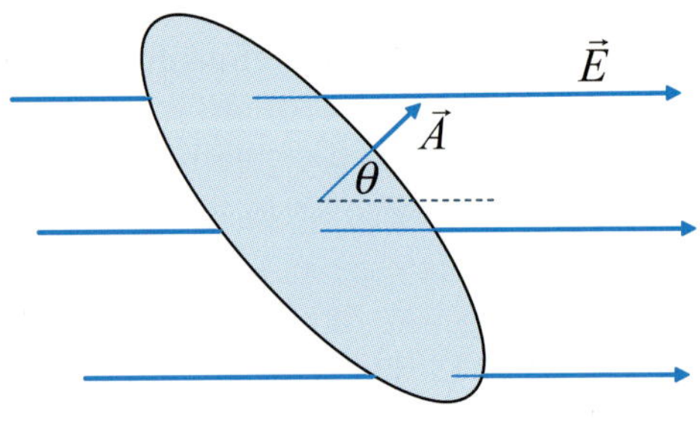

$$\Phi = (E\cos\theta)\,A = \vec{E}\cdot\vec{A}$$

Figure 1.17

The **flux** Φ of the electric field through this area is the magnitude of the area multiplied by the component of the field that is perpendicular to the area. Thus

$$\Phi = (E\cos\theta)A = \vec{E}\cdot\vec{A}$$

This equality expresses the flux as a dot product of the two vectors. To find flux for a nonuniform field or through a more contorted area, you would have to break up the surface into many different, smaller patches, such as the two shown in Figure 1.16, $\Delta\vec{A}_{1i}$ and $\Delta\vec{A}_i$. The units of the electric flux are $\text{N}\cdot\text{m}^2/\text{C}$.

Note that there is some ambiguity in our definition of flux. Since there are two ways to be perpendicular to an area, the area vector could have pointed in the opposite direction. In any case, for a closed surface, there is no ambiguity about the direction of the area vector: it is perpendicular to the area patch and toward the *outside* of the surface, as shown in Figure 1.18a.

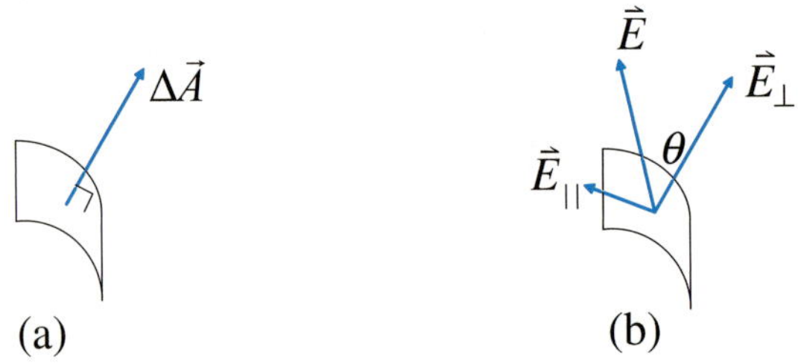

Figures 1.18a and 1.18b

To determine flux, then, you sum up all contributions from the little patches, taking the limit where the patches become infinitesimally small: $\Delta\vec{A} \rightarrow d\vec{A}$. So the sum becomes an integral and you have

$$\Phi = \oint_{surface} \vec{E} \cdot d\vec{A}$$

As you can see from Figure 1.18b, only the perpendicular components of $\vec{E}$ will contribute to this integral, because only these components actually cut through the area.

For a sphere, as in Figure 1.16, the electric field is always parallel to the area vectors on the surface. Thus, for every patch on the sphere

$$\vec{E} \cdot d\vec{A} = EdA$$

Since the magnitude of $\vec{E}$ is constant over the sphere, it factors out of the integral and you have

$$\Phi = \oint_{surface} \vec{E} \cdot d\vec{A} = E \oint_{sphere} dA = E4\pi r_1^{\,2}$$

Substituting for E, you get

$$\frac{1}{4\pi\varepsilon_0}\frac{q}{r_1^{\,2}}(4\pi r_1^2) = \frac{q}{\varepsilon_0}$$

Thus the flux through the surface is proportional to the charge inside the surface. The flux does not depend at all on the radius of the sphere as long as the sphere encloses the charge. Because the field decreases as $\frac{1}{r^2}$ and the area increases as r^2, the two factors are exactly equivalent to a factor of 1—they "cancel out." This is just what happened with the photons: while the intensity of the light was different at the different surfaces, the total number of photons per second passing through any closed surface stayed the same. For a given charge, the strength of the field is different on different spheres surrounding the charge, but the flux is the same through all of them.

If the surface enclosing the charge has an odd shape like the potato skin, the electric flux is still the same. The argument is exactly as it was for the photons. The field may cut through larger areas, such as $\Delta\vec{A}_i$ in Figure 1.16, but these areas will be oriented so that the field is not perpendicular to the area. The overall effect is that the flux through the inner sphere will be the same as through the potato skin. As long as the area encloses the charge, you can be sure

$$\Phi = \oint_{\substack{closed\\surface}} \vec{E} \cdot d\vec{A} = \frac{q_{inside}}{\varepsilon_0}$$

If the surface encloses more that one charge, you can use the principle of superposition for electric fields. The total field is the vector sum of the fields due to the individual charges. You can then argue that the flux, a scalar, is the sum of the individual fluxes. The charge inside becomes the algebraic sum of the individual charges.

Even with these expanded meanings for $\vec{E}$ and q_{inside}, you can write Gauss's Law in its final form as

$$\oint_{\substack{closed\\surface}} \vec{E} \cdot d\vec{A} = \frac{q_{inside}}{\varepsilon_0}$$

This result is truly remarkable. No matter how contorted you make the surface, no matter how many charges happen to be present, the flux of the field through any closed surface is directly proportional to the charges enclosed within the surface. This fundamental result is a direct consequence of the dependence of the field of a point charge on the ratio $\frac{1}{r^2}$ and of the superposition properties that the laws of electricity obey.

Exercises

1. In Figures 1.1, 1.2, and 1.3 of this chapter and the surrounding text, the process of charging a conducting ball negatively by induction was described. Explain how to charge the conducting ball positively by induction, drawing the corresponding figures.

2. Sketch the electric field pattern for the region around two negative point charges of equal magnitude.

3. A circular nonconducting hoop of radius R is in the region near a large nonconducting plate. The surface charge density on the plate is uniform and equal to σ. Find the flux through the hoop for each of the three positions shown in Figure 1.19. Express your answer in terms of σ, R, and fundamental constants.

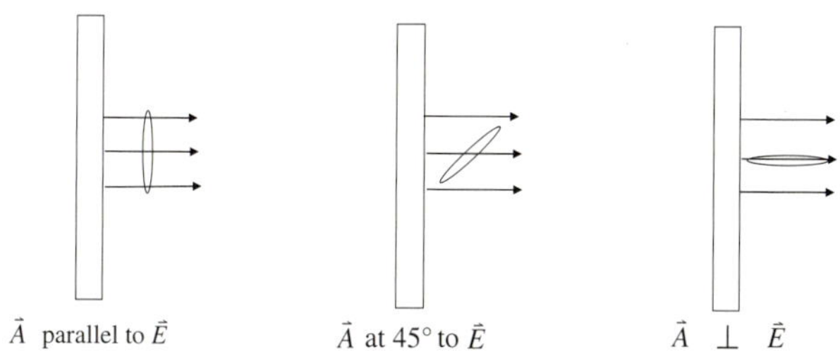

Figure 1.19

4. A point charge of magnitude 2.0 μC (2.0×10^{-6}C) is at the center of a cubical Gaussian surface. The area of each face of the cube is 9.0×10^{-2} m^2. What is the net electric flux through the surface of the cube shown in Figure 1.20?

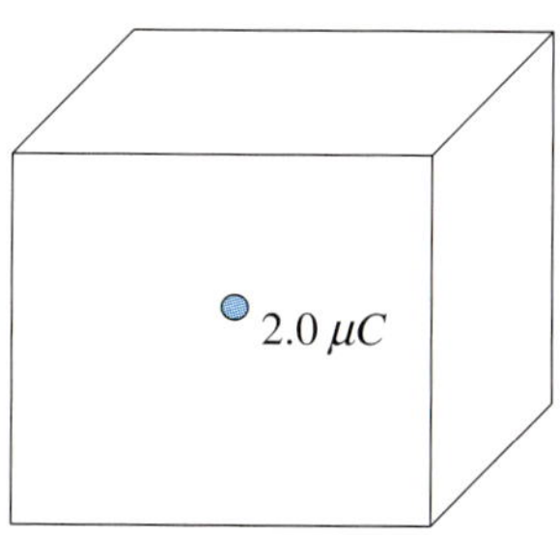

Figure 1.20

5. For the same charge as in Exercise 4, remove all but one face of the cube. What is the flux through the remaining plane surface shown in Figure 1.21?

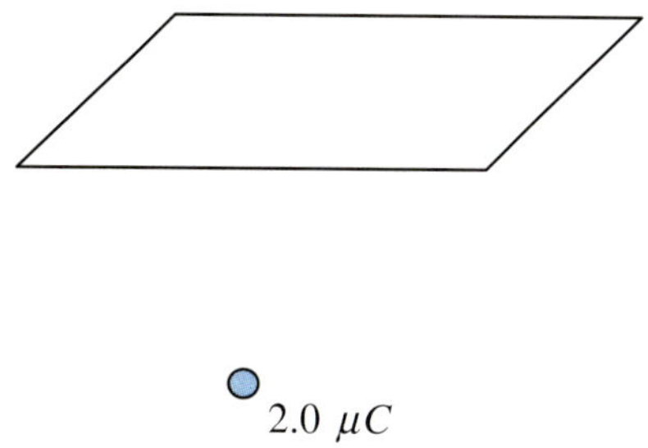

Figure 1.21

6. Find the flux through the odd-shaped surface with cross section shown in Figure 1.22.

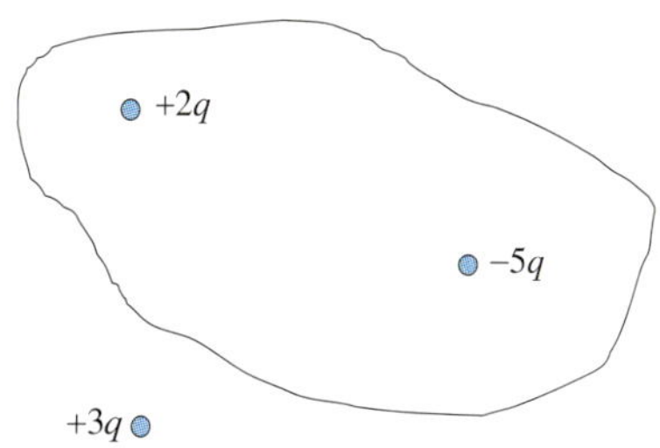

Figure 1.22

7. If $E = 0$ everywhere on a closed surface, is it correct to say that there is no net charge inside the surface? Explain.

8. If there is no net charge inside a closed surface, is it correct to say that $E = 0$ everywhere on the surface? Explain.

Problems

1. Two positive point charges of equal magnitude q are located at the coordinates $\left(\frac{-a}{2}, 0\right)$ and $\left(\frac{+a}{2}, 0\right)$, as shown in Figure 1.23.

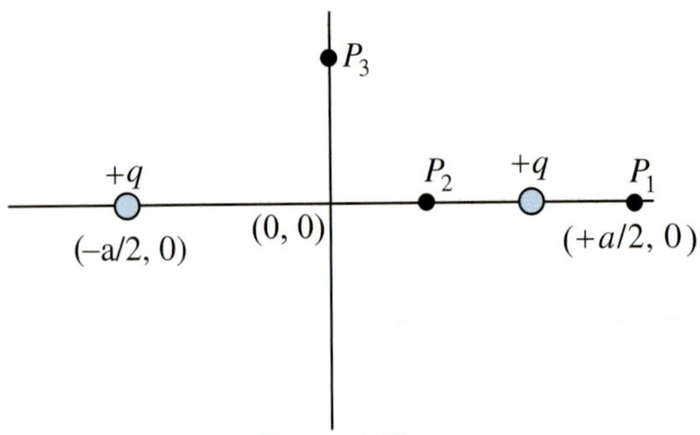

Figure 1.23

Find a vector expression for the electric field at

a. a point P_1 on the x-axis for $x > a/2$, in terms of x, a, and fundamental constants. When $x >> a/2$, what value does the field approach? Does this seem reasonable?

b. a point P_2 on the x-axis for $0 \leq x < a/2$, in terms of x, a, and fundamental constants. Does this answer seem reasonable when $x = 0$?

c. a point P_3 on the y-axis, in terms of y, a, and fundamental constants. When $y >> a/2$, what value does the field approach? Does this seem reasonable?

2. A small piece of polystyrene foam with mass m and charge q is at rest in a vacuum above a large nonconducting plate with uniform surface charge density σ of the same sign as q.

a. Draw a free body force diagram for the piece of polystyrene.

b. Solve for the surface charge density on the plate in terms of m, q, and fundamental constants.

3. A line of charge of constant linear charge density $+\lambda$ and length L occupies the part of the x-axis where $x \geq a$, as shown in Figure 1.24. Find the electric field strength at the origin.

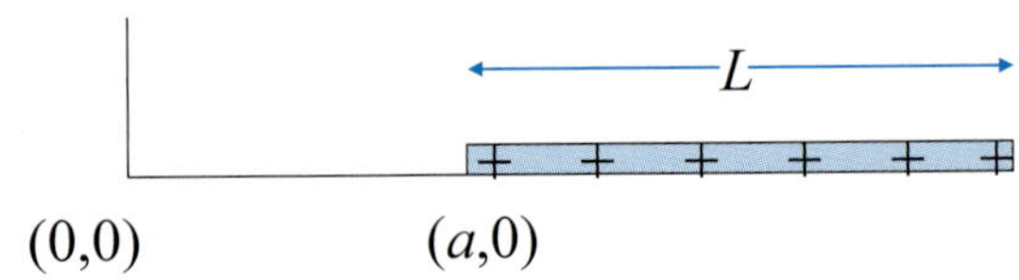

Figure 1.24

CHAPTER 2

Applications of Gauss's Law

2.1 General Applications of Gauss's Law

Refer back to Figure 1.7, and imagine that a closed surface surrounds the pair of equal and opposite charges illustrated there. The situation might look as in Figure 2.1 below. Such a closed surface, indicated by the dotted line, is often called a **Gaussian surface**.

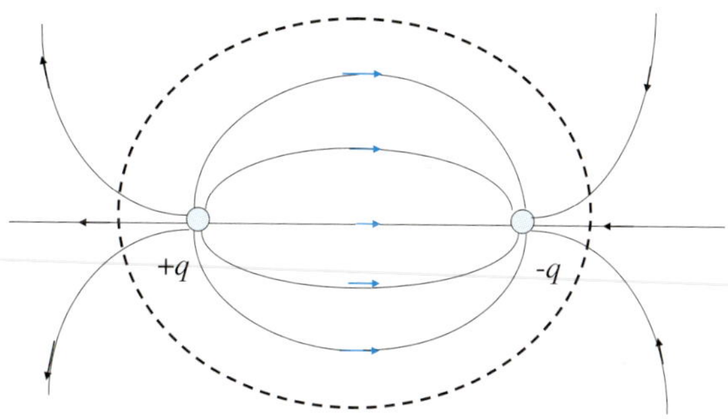

Figure 2.1

Gaussian surfaces are usually make-believe surfaces that are introduced in the analysis of a problem in electrostatics. Their sole purpose is to provide the closed surface necessary for the application of Gauss's Law. Choosing the right surface to make a clear analysis of a particular problem is a talent you will have to develop. The flux through the surface is indicated by the number of lines going through the surface. In the example given above, three lines leave the surface and three lines enter the surface. The net flux is zero, and the net charge contained inside the surface is also zero. If you take the same diagram, but this time draw a Gaussian surface enclosing only one of the charges, it would look like Figure 2.2.

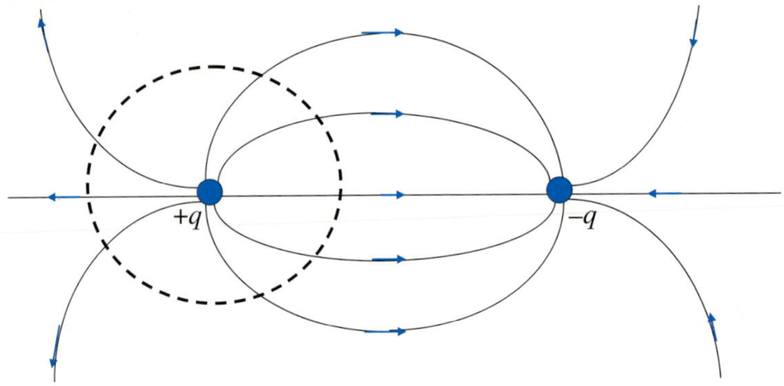

Figure 2.2

Here eight lines leave and none enters. The eight lines represent the flux due to the charge $+q$. Any Gaussian surface surrounding only $+q$ and not $-q$ would have the same eight lines leaving it. Flux is proportional to the charge enclosed.

Try this same exercise again, now referring back to Figure 1.8. Draw a surface enclosing both charges given there. One possibility is shown in Figure 2.3.

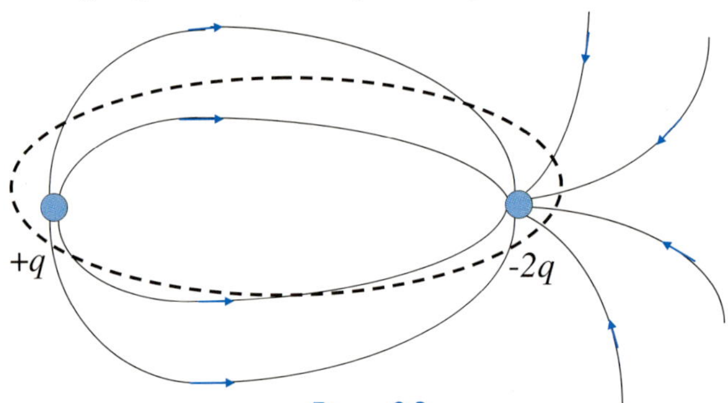

Figure 2.3

In this instance of enclosing the charges, three lines leave and seven lines enter the surface, so the net flux, $3 - 7$, is -4. Note that 4 is the number of lines leaving $+q$ and the number of lines representing a charge of magnitude q. But the net charge enclosed within the Gaussian surface is $-q$. So again, flux is proportional to the charge enclosed.

2.2 Applications to Conductors

Imagine that you charge up a solid conductor of any shape negatively. You give the extra electrons time to come to rest in their equilibrium positions, and the charge is then said to be **static**. This means that the electric field inside the conductor must be zero—otherwise the charges would respond to the field and move. However, you might wonder: Could the excess charge be somehow spread out inside the conductor? If this were possible, and you were to draw a Gaussian surface just inside the outer surface of the conductor, as shown in Figure 2.4a, there would be charge inside this surface. That means there would be flux through the surface, but there can be no flux if there is no field. Therefore, you can see that the extra electrons depicted in Figure 2.4b must have come to rest on the *outside* surface of the conductor.

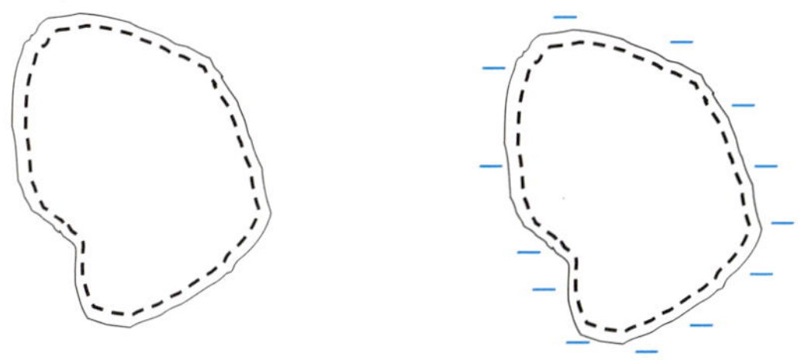

Figures 2.4a and 2.4b

For a positive net charge on the conductor, the argument is the same. The excess positive charge will also reside only on the exterior surface. (The implication is not that electric fields can never exist within conductors. When you use a light bulb or a toaster, there are electric fields in the wires that let the current flow; however, here we are discussing only static electricity.)

The same result will hold for a hollow conductor, sometimes called a **conducting shell**. Consider a Gaussian surface inside the body of a hollow conducting shell like the one shown in Figure 2.5a. We have already established that there can be no field inside the body of a conductor. If there is no field, there can be no flux through this Gaussian surface, and no net charge within that surface. The excess charge must be on the outside surface of the conducting shell. Now imagine another Gaussian surface in the empty space inside the shell (Figure 2.5b). It has no charge inside it; therefore, there is no flux through it and no field in this space.

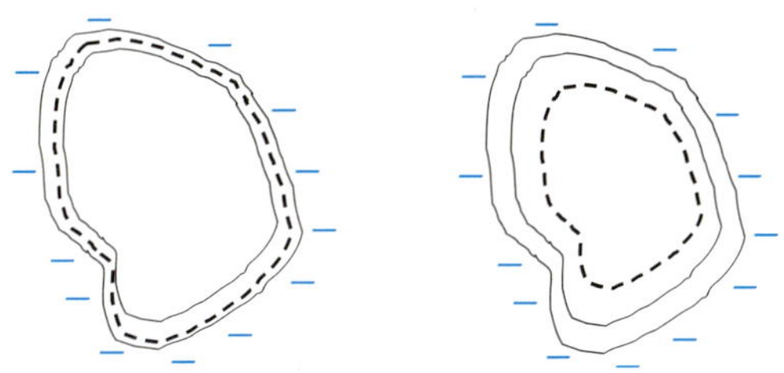

Figures 2.5a and 2.5b

You can also understand this result qualitatively from studying Figure 2.5b, if you think about a small positive charge located at any point on the Gaussian surface. This charge will be pushed to the right by the charges on its left, and vice versa, so that all of these forces cancel out. Moreover, if there were to be a field, the field lines would have to terminate on the negative charges. But where would the lines start? There is no charge inside the Gaussian surface to act as starting points for the field lines.

The fact that the electric field is zero inside a hollow conducting surface is important because it allows you to protect something—for example, sensitive electronic equipment—from the influence of outside charges or fields simply by surrounding the equipment by a closed metal box. Another important application is the Van de Graaff generator. Charge applied to the inside of a large metal sphere will not stay there. The charge migrates to the outside surface, where it can be stored until it discharges in a large spark.

It is also important to remember that the electric field at the surface of the conductor must be perpendicular to the surface. If it were to have any parallel component, then the excess charge would respond to that component by moving around on the surface. Because the situation is static, this cannot happen.

To sum up our conclusions:

- There can be no electric fields inside charged static conductors.
- The charge resides on the exterior surface of the conductor.
- The electric field at the surface of a conductor is perpendicular to the surface.

2.3 Spherical Symmetry

Now let's tackle some of the many problems to which Gauss's Law applies.

Example 1 A small conducting sphere of radius a has a charge $+q$. Find expressions for the electric field strength

a. inside the sphere

b. outside the sphere

Solutions

a. As you will recall by looking back at Figure 2.4, we have established that the field inside a charged conductor must be zero.

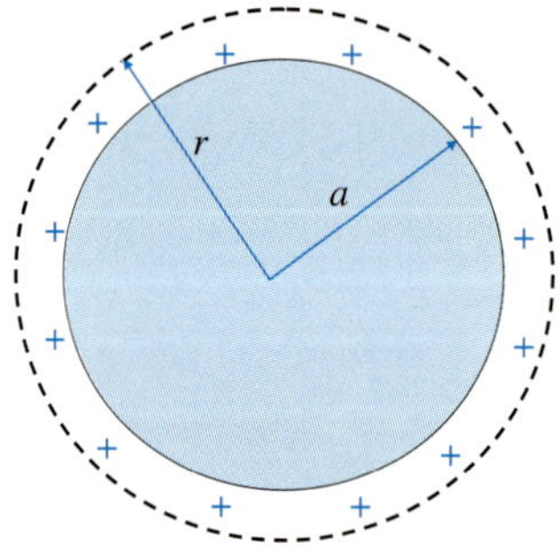

Figure 2.6

b. For the field outside the sphere, make a sketch that includes a Gaussian surface at a radius r greater than that of the conducting sphere, as shown in Figure 2.6. We have chosen the Gaussian surface to be spherical because the electric field and area vectors are radial. You know that the area vectors point outward along radii everywhere on the sphere. You also know, by symmetry, that the electric field vectors must be radial too. Thus, you may write the flux through the area dA as

$$\vec{E} \cdot d\vec{A} = EdA$$

and restate the problem as one of finding the electric field at any point on our Gaussian surface—that is, at the radius r. Now, write down Gauss's Law.

$$\oint_{\substack{closed\\surface}} \vec{E} \cdot d\vec{A} = \frac{q_{inside}}{\varepsilon_0}$$

Symmetry also tells you that E must have the same value at every point on the sphere of radius r. This means that E factors out of the integral, so we have

$$\oint_{sphere} \vec{E} \cdot d\vec{A} = \oint_{sphere} E dA = E \oint_{sphere} dA = E(4\pi r^2)$$

given the area of a sphere, $A = 4\pi r^2$. Whenever Gauss's Law is properly applied to a spherically symmetric problem, the flux of the electric field through the Gaussian surface always reduces to $E(4\pi r^2)$. So in this case, Gauss's Law becomes

$$E(4\pi r^2) = \frac{q}{\varepsilon_0}$$

$$E = \frac{1}{4\pi\varepsilon_0} \frac{q}{r^2}$$

Note that this is the same expression as for the magnitude of the field strength due to a point charge (see Section 1.2).

What about direction of the field? We have a positive charge on our conducting sphere. So the electric field vectors must point radially outward, as illustrated in Figure 2.7.

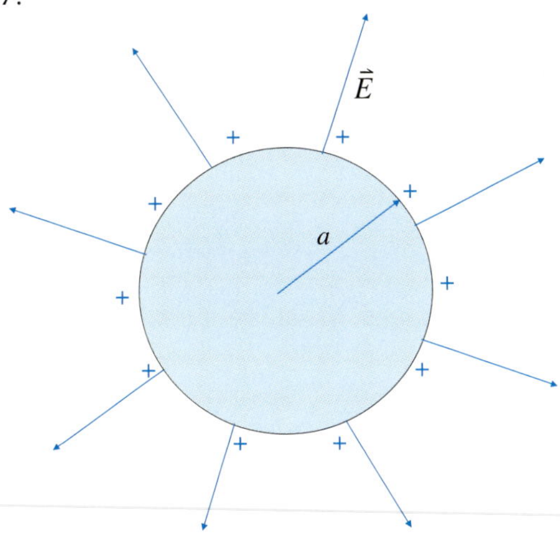

Figure 2.7

Example 2 A small conducting sphere of radius a has a charge $+q$. It is surrounded by a spherical conducting shell of inner radius b and outer radius c. The shell has a charge $-q$. The general problem is to determine the electric field at various radii r from the center of the spheres. Find expressions for the magnitude of the electric field where

a. $r < a$

b. $a < r < b$

c. $b < r < c$

d. $r > c$

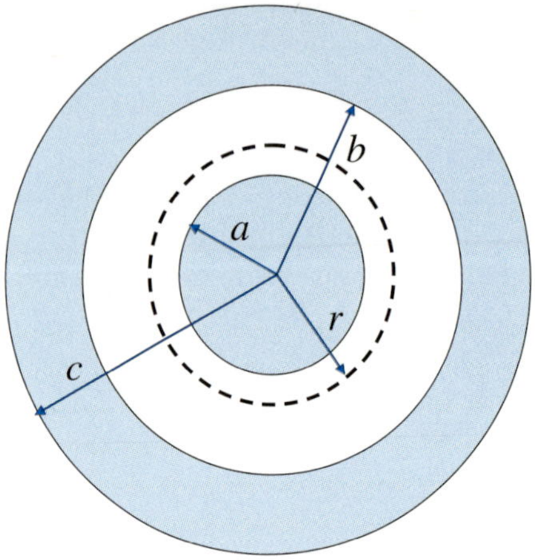

Figure 2.8

Solutions

For parts *a* and *b*, the problem situation is illustrated by Figure 2.8, where you also see a spherical Gaussian surface of radius *r* drawn in the space between the inner sphere and the shell.

a. As in the previous example, inside a conductor $E = 0$.

b. The mathematics here is the same as for the second part of Example 1. The problem has spherical symmetry, so again the left side of the Gauss's Law equation will be $E(4\pi r^2)$. Only the charge on the inner sphere will be included in the Gaussian surface. The result is

$$E = \frac{1}{4\pi\varepsilon_0}\frac{q}{r^2}$$

The inner sphere acts like a point charge for points outside. Only the charge on the inner sphere determines the field in the gap. The charge on the outer shell makes no contribution.

c. Now refer to Figure 2.9. At this radius you are again inside the body of a piece of metal, so $E = 0$. An interesting point arises here. If $E = 0$, then Gauss's Law says that there must be no charge inside that Gaussian surface. This means that the charge $-q$ on the shell must reside on the inner surface of the shell. The positive charge $+q$ on the inner sphere attracts just enough negative charge inward on the outer shell to insure that the field inside the shell is zero.

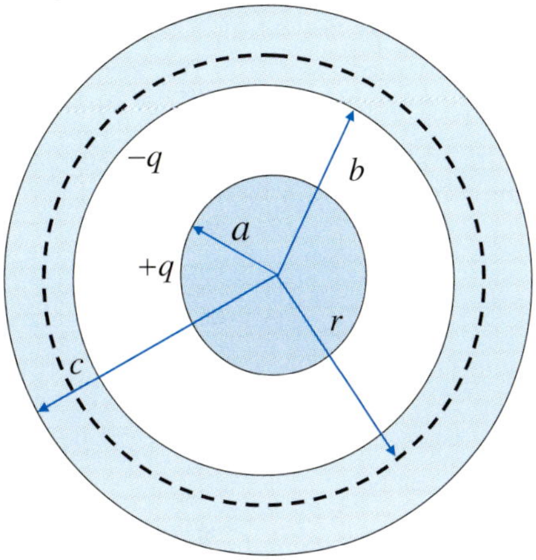

Figure 2.9

d. In this case, for $r > c$, a dotted sphere denoting the Gaussian surface will enclose the entire diagram. There is no net charge enclosed, so E again must be zero.

The device illustrated in this example is called a **spherical capacitor**. Charge can be stored on it and the electric field is confined to the space between the inner and outer charged surfaces. A diagram of the electric field vectors is given in Figure 2.10.

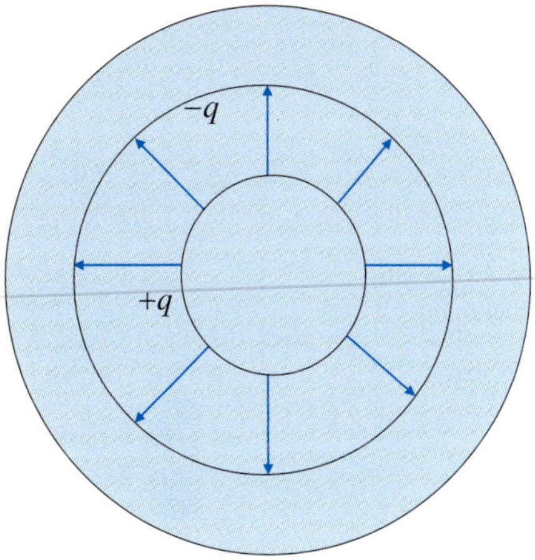

Figure 2.10

In such a device, how is it possible to get equal and opposite charges on the inner sphere and the outer shell? You certainly could not do it by touching each of them in turn with charging strips. Here is a way to do it that relies on Gauss's Law for its success. First, charge up the inner sphere. Next, bring in two uncharged hemispherical shells and hook them up to form a spherical shell concentric with the first sphere. Electrons on the shell will move to the inner surface because they are attracted by the positive charge; just the right number will move in to satisfy Gauss's Law and leave the field in the body of the shell zero. This leaves the outer surface of the shell positive by the same amount. Nothing has touched the shell, so it cannot have a net excess charge. Finally, attach the outer shell to ground. Electrons will come up from ground to cancel out the excess positive charge on the exterior surface of the shell. The process is shown in Figures 2.11 and 2.12.

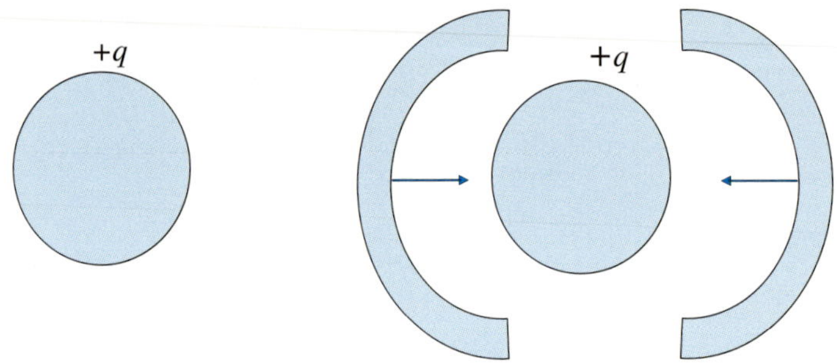

Figure 2.11

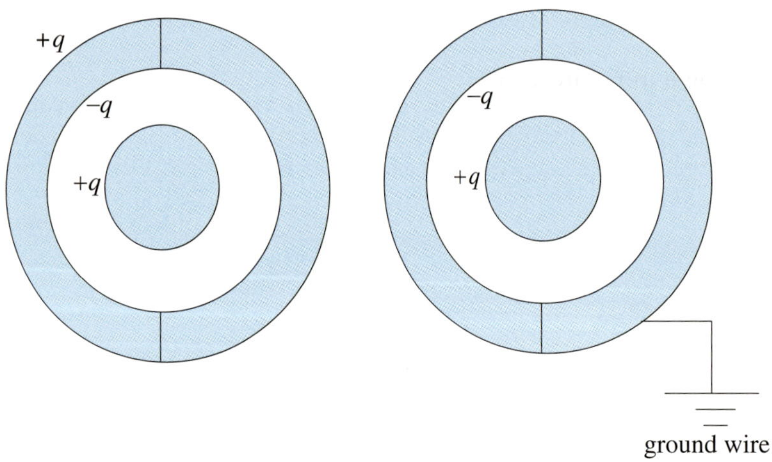

Figure 2.12

If a capacitor of any shape is filled with a dielectric material, instead of having a vacuum between the charged surfaces, the expressions for E are the same, except that ε_0 will be replaced by ε. The value of ε depends on the material. You may wonder how the field strengths would work out if there were unequal charges on the sphere and the shell. To find out, let's do another example.

Example 3 Refer back to Figure 2.8, but this time assume that the inner sphere has charge $+3q$ and the outer sphere has charge $+2q$. Find expressions for the electric field strength at various radii r where

a. $r < a$

b. $a < r < b$

c. $b < r < c$

d. $r > c$

Solutions

a. As in all cases inside a conductor, $E = 0$.

b. The charge enclosed is the $+3q$ on the inner sphere. For points outside of this sphere, this charge behaves as a point charge, so you have

$$E = \frac{1}{4\pi\varepsilon_0}\frac{3q}{r^2}$$

c. $E = 0$ inside the metal of the conducting shell.

d. The net charge enclosed is $+5q$, which behaves as a point charge for points outside.

$$E = \frac{1}{4\pi\varepsilon_0}\frac{5q}{r^2}$$

Gauss's Law tells us that, if the field inside the conducting shell (part **c**) is zero, there must be a charge of $-3q$ on the inner surface of the shell. Because the net charge on the shell is $+2q$, the charge on the outer surface of the shell must be $+5q$. The presence of the $+3q$ on the inner sphere has caused charge to separate on the shell in just the right amount to ensure that the field inside the body of the conducting shell is zero. The overall picture is shown in Figure 2.13.

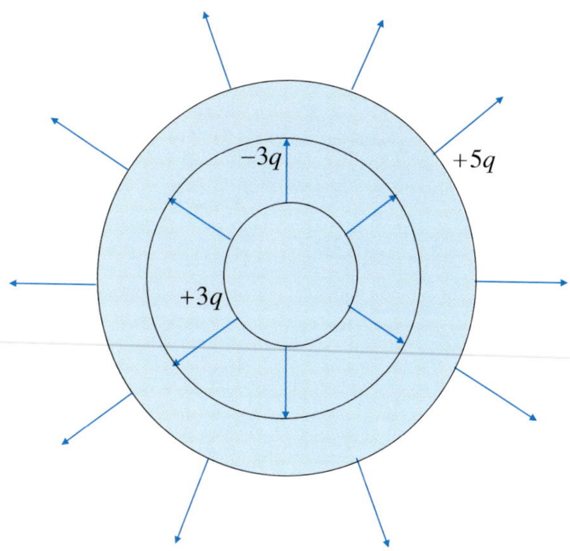

Figure 2.13

Also note that the resulting field expression for $r > c$ (part **d**) shows that the field outside the shell is determined only by the charge on the outer surface of the shell. Accordingly, for every three lines in the gap, there are five lines outside.

2.4 Planar Symmetry

In Example 2 of Chapter 1, you may recall the solution to the problem of finding an expression for the electric field near a large, flat, charged nonconducting plate. At the end of that example, you were promised that Gauss's Law could provide a simpler solution to the same problem. Now is the time to deliver on that promise!

Example 4

Figure 2.14 shows a large, flat nonconducting plate with charge per area σ spread uniformly all over it. Find the expression for the field at a point near the surface of the plate.

Solution

One challenge of applying Gauss's Law is deciding on the shape of the Gaussian surface that will simplify the integral. To make a good choice, you have to know something about the direction of the field. In Example 2 of Chapter 1, assuming you were not too near the edge of the plate, you saw that the vector fields from the charges on the plate formed a cone such that the components of the field parallel to the plate cancelled out (see Figure 1.12). $\vec{E}$ was perpendicular to the plate. You must use this same reasoning in this alternate solution to the problem. Consider the Gaussian surface shown in Figure 2.14, extending equal distances above and below the plate. It has the advantage that, along the sides of the cylinder, $\vec{E}$ is parallel to the surface, and hence perpendicular to the vector $\vec{A}$.

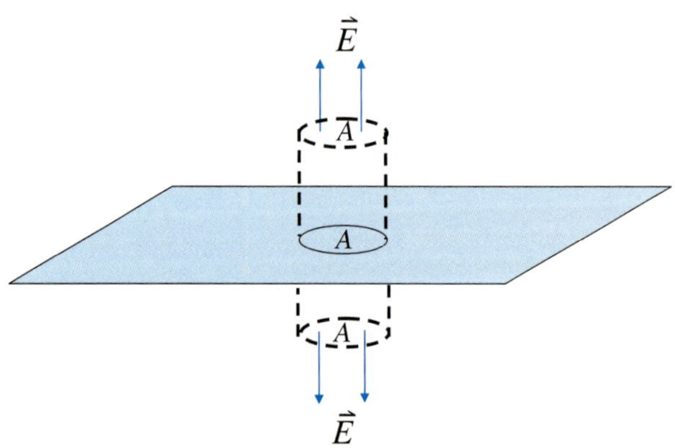

Figure 2.14

The dot products $\vec{E} \cdot d\vec{A}$ will be zero on the curved surface, or sheath, of the cylinder. This leaves only the bases of the cylinder to consider. There $\vec{E}$ is constant, perpendicular to the plate and parallel to the vector $d\vec{A}$ elements. These observations ensure that the flux through this surface is

$$\oint_{cylinder} \vec{E} \cdot d\vec{A} = 2 \int_{base} E dA = 2E \int_{base} dA = 2EA$$

Note that the flux is positive through both bases. The field is in opposite directions at the two bases, but the area vector, directed along the outward normal, is also in opposite directions at the two bases. Applying Gauss's Law to this cylinder you get

$$2EA = \frac{q_{inside}}{\varepsilon_0}$$

You must still consider how to find the charge inside. Because the cylinder encloses an area A of the plate, the charge inside the Gaussian surface is $q = \sigma A$. Substituting into Gauss's Law yields

$$2EA = \frac{\sigma A}{\varepsilon_0}$$

$$E = \frac{\sigma}{2\varepsilon_0}$$

This result is the same answer you have seen for this problem in Chapter 1. Recall that $\bar{E}$ is independent of the distance from the plate as long as the point remains close enough to the plate and is not too close to the edge.

Example 5 Sketch the general shape of the electric field in the region between a point charge $+q$ and a conducting plate with no excess charge on it.

Solution This problem can be tricky because the plate is a conductor. Charges may become separated on the two sides of the plate, even though the overall charge on the plate is zero. This phenomenon is similar to what happened with the conducting shells in Examples 2 and 3. Here, recall that the field due to a point charge alone is a set of radial lines, and that the field at the surface of a conductor is perpendicular to the surface. Using these facts, we can make a sketch of the field like the one in Figure 2.15.

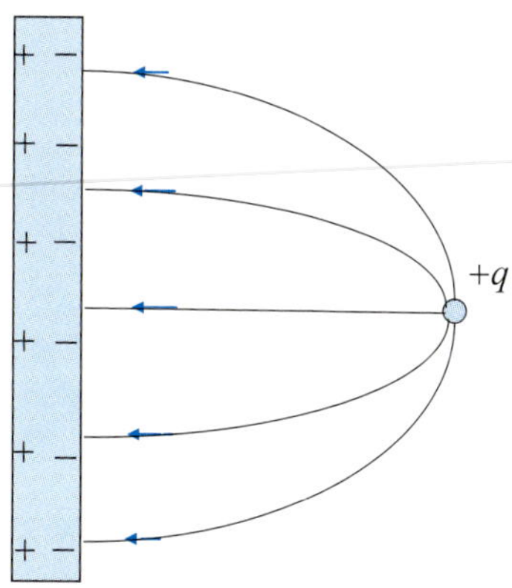

Figure 2.15

The plate alone produces no electric field, because it carries no excess charge. The point charge by itself produces radial field lines that are centered on the point charge. However, the combination of the plate and the point charge produces a field that is not the simple superposition of the fields of the two individual components. This is because the presence of the point charge induces a charge distribution on the conductor. When conductors are present, you have to know how the charge redistributes before you can effectively apply Gauss's Law.

Example 6

Figure 1.9 of Chapter 1 shows an arrangement of two large, parallel conducting plates, with equal and opposite charges on each plate. The mutually attractive forces between the opposite charges have drawn all the excess charge to the interior surfaces of the plates. The surface charge density σ is uniform over each plate. Calculate the field

a. between the plates

b. outside the plates

Solutions

Because you are given the equilibrium charge distribution on the conductors, you can apply Gauss's law easily.

a. In Figure 2.16 the plates are seen edge on, and the dotted rectangle is the edge-on view of a Gaussian cylinder.

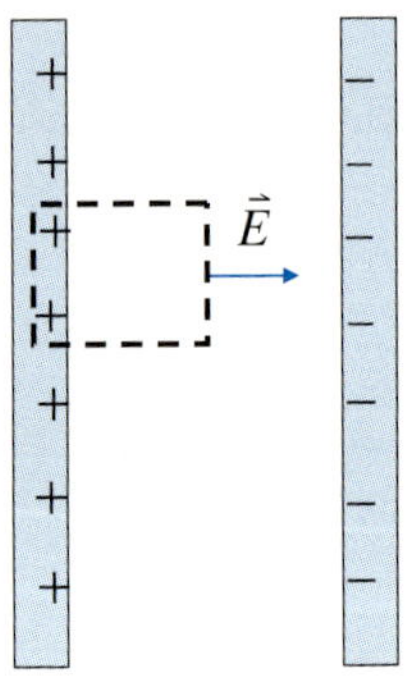

Figure 2.16

Symmetry tells you that the field between the plates must be perpendicular to the plates themselves. The field on the left end of the cylinder must be zero because that surface is located inside a conductor. Applying Gauss's Law, you get

$$\oint_{\substack{closed \\ surface}} \vec{E} \cdot d\vec{A} = \frac{q_{inside}}{\varepsilon_0}$$

$$EA = \frac{\sigma A}{\varepsilon_0}$$

$$E = \frac{\sigma}{\varepsilon_0}$$

b. To find the field outside the plates, use the Gaussian surface shown in Figure 2.17.

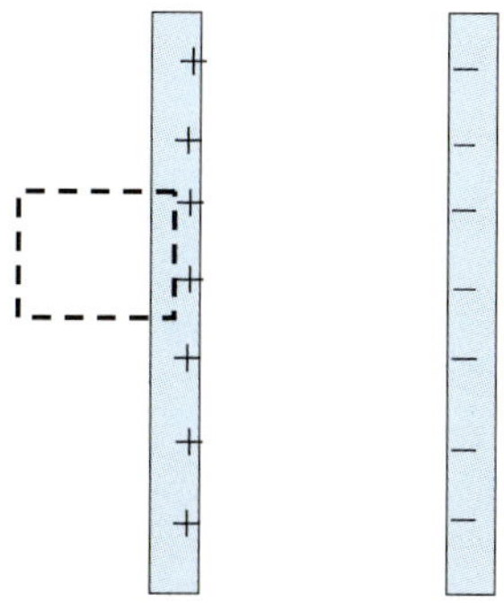

Figure 2.17

Because all the charge is on the inner surface, there is no charge enclosed by the Gaussian surface. There is no field in the conductor, so the flux through the base of the cylinder inside the conductor is zero. If there were to be any field on the left side of the Gaussian cylinder, it would have to be perpendicular to the plates and would produce a flux through the left base of the cylinder. But, with no charge enclosed, there cannot be any such field. Therefore, $E = 0$ outside the plates.

Parallel, oppositely charged plates have some important practical uses. In TV sets, computer monitors and oscilloscopes where a cathode ray tube is used, parallel plates may be used to deflect the electron beam to different parts of the screen. There are usually two sets of plates, one for horizontal and one for vertical deflection. When the voltages on the plates vary, the charge densities change also, and the electron beam can be deflected more or less accordingly. A set of oppositely charged parallel plates is also the simplest form of capacitor, a device for storing charge.

2.5 Cylindrical Symmetry

Problems with cylindrical symmetry are another group of problems made simpler by Gauss's Law. The simplest of these is a long line of charge. In Example 7 and the cylindrical examples to follow, true cylindrical symmetry is achieved only if the cylindrical charge distributions are infinite in length. However, a very good approximation will result from using cylindrical symmetry and applying Gauss's Law when the distance r from the axis of the cylinder is much less than the length L of the charge distribution.

Example 7 Charge Q is spread uniformly on a long line of length ℓ. Find the magnitude of the electric field at a point P a distance r from the line.

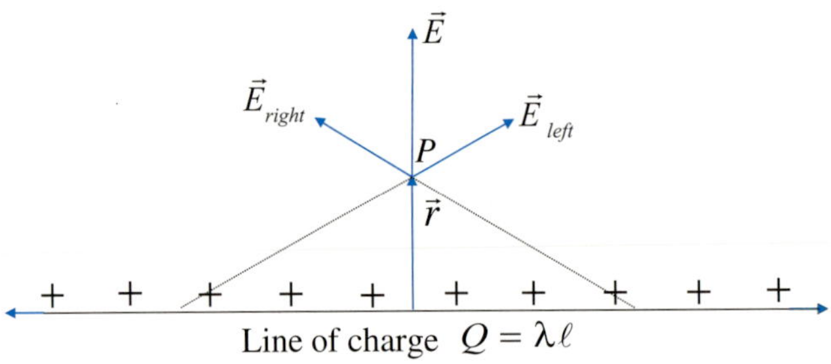

Horizontal contributions from symmetric bits of charge on either side of *P* will cancel.

Figure 2.18

Solution

Just as in the examples involving planar symmetry, you first have to determine the direction of the field in order to choose a useful Gaussian surface. If you are not too close to either end of the line, the vector fields from each little part of the line combine to give a net field that points directly away from the line. In Figure 2.18, you can see that the horizontal components from symmetric elements on the left and right will cancel out, leaving only the components of $\vec{E}$ perpendicular to the line. Let's define a quantity called the **linear charge density**—the charge per unit length. Its symbol is λ and its units are coulombs per meter.

$$\lambda = \frac{Q}{\ell}$$

You can see that, in terms of λ, the charge on an arbitrary length L of the line will be $q = \lambda L$.

Now you need to choose a Gaussian surface. A good choice is the cylinder shown by the dotted line in Figure 2.19. The cylinder is chosen to have radius r, because that is the distance from the axis at which you want to calculate the field. The length L of the cylinder can be any value and, because this is true, you should expect the result for the field to be independent of L.

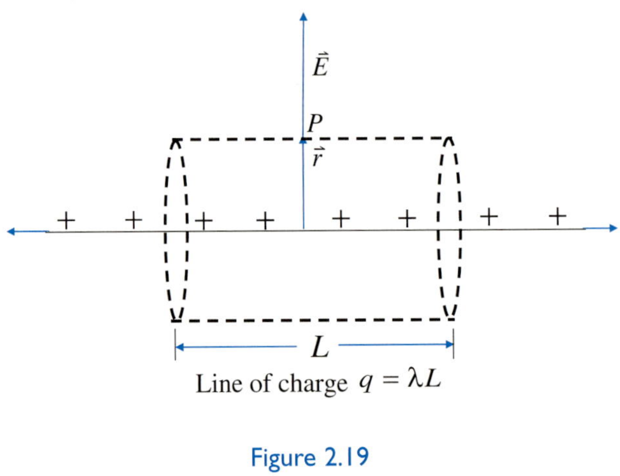

Figure 2.19

The electric field vectors will be perpendicular to the sheath or curved surface of the cylinder and parallel to its bases. There is flux only through the sheath and not through the bases. Write down Gauss's Law:

$$\oint_{\substack{closed\\surface}} \vec{E} \cdot d\vec{A} = \frac{q_{inside}}{\varepsilon_0}$$

The dot product $\vec{E} \cdot d\vec{A}$ becomes EdA over the sheath and zero over the bases. Over the sheath, the field may be factored out of the integral because it has a constant magnitude everywhere on the sheath. The charge q may be replaced by λL. This gives

$$E \oint_{\substack{closed\\surface}} dA = \frac{\lambda L}{\varepsilon_0}$$

So the integral is just the surface area of the sheath of the cylinder. Now imagine a hollow cylinder made out of a piece of paper. If you open it up along its length L and flatten it out, it will look like a rectangle. The length of the rectangle will be L and its width will be the circumference of the circular base of the cylinder, namely, $2\pi r$. The area of the rectangle is, therefore, $2\pi rL$. Substituting,

$$E 2\pi rL = \frac{\lambda L}{\varepsilon_0}$$

Note that the flux integral on the left-hand side of Gauss's equation has reduced to $E(2\pi rL)$. This always happens in problems with cylindrical symmetry if you draw your Gaussian surface concentric to the symmetry axis. Solving for E gives

$$E = \frac{\lambda}{2\pi\varepsilon_0 r}$$

The magnitude of the electric field varies inversely as the distance r from the line and is directly proportional to the linear charge density λ. The field does not depend on the length of the cylinder chosen for the Gaussian surface, as you might expect.

Example 8 A long cylinder of radius a and length L is covered uniformly with charge on its curved surface. The linear charge density is λ. Find expressions for the electric field where

a. $r < a$

b. $a < r << L$

Solutions

a. First, sketch a picture of the situation, something like Figure 2.20. The line in the center is the axis of the cylinder, from which all radii may be measured. In addition, a Gaussian surface, a cylinder of radius r, needs to be drawn in. Because there is no charge inside the Gaussian surface, for any radius less than a, $E = 0$.

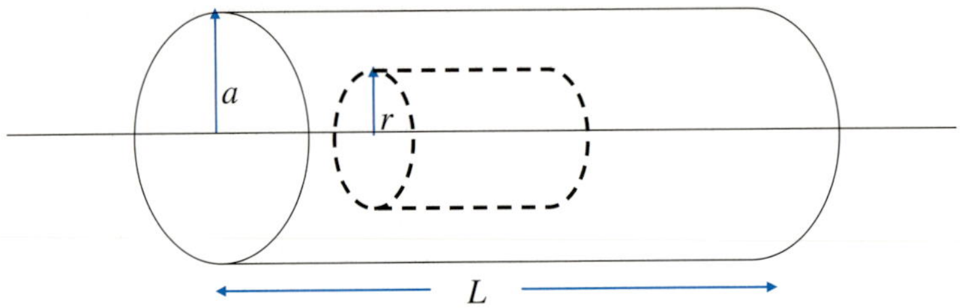

Figure 2.20

b. For this restriction on r, redo the sketch, this time with a Gaussian surface as shown in Figure 2.21.

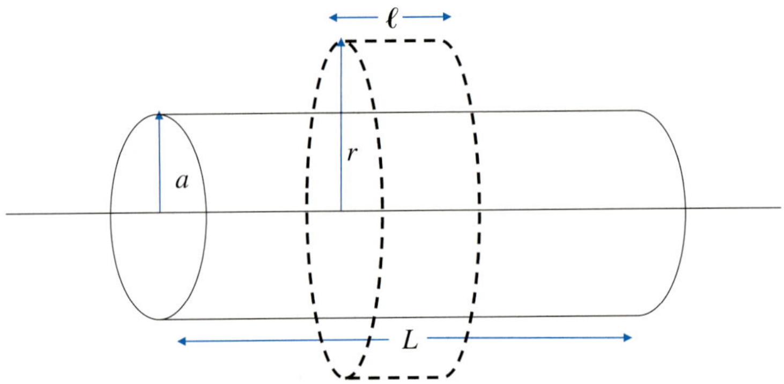

Figure 2.21

Start with Gauss's Law:

$$\oint_{\substack{closed\\surface}} \vec{E} \cdot d\vec{A} = \frac{q_{inside}}{\varepsilon_0}$$

Just as in Example 7, the electric field vector on the curved part of the Gaussian surface is perpendicular to the axis of the cylinder, provided $r << L$. The field is also constant everywhere on that surface. You can easily see that the equation becomes

$$E2\pi r\ell = \frac{\lambda\ell}{\varepsilon_0}$$

$$E = \frac{\lambda}{2\pi\varepsilon_0 r}$$

This is the same result as for the line of charge. When you are inside the cylinder, the field is zero; when you are outside the cylinder, the cylinder behaves as a line of charge corresponding with the axis. Do you see the clear parallel to the spherically symmetric examples in Section 2.3?

Let's continue to look at the parallel with cases of spherical symmetry, and solve a problem involving two concentric charged cylinders.

Example 9 A solid conducting cylinder of radius a is covered with charge $+q$ spread uniformly over it. The charge equals λL, where L is the length of the cylinder. There is also a concentric, thin cylindrical shell of radius b. The shell has the same length as the inner cylinder and an equal and opposite charge $-q$. Find expressions for the electric field for

a. $r < a$

b. $a < r < b$

c. $b < r << L$

Solutions

a. There is no charge inside a Gaussian cylinder with radius $r < a$, where a is the radius of the conducting cylinder. $E = 0$.

b. For $a < r < b$, imagine a Gaussian cylinder of an intermediate radius r and length ℓ. This case is represented by Figure 2.22.

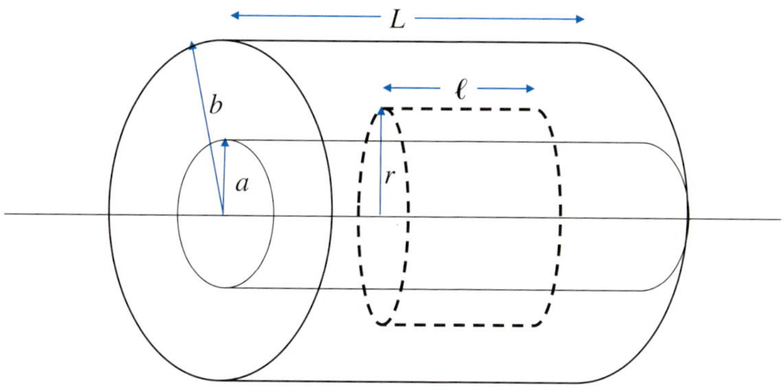

Figure 2.22

The left side of Gauss's equation becomes $E2\pi r\ell$. The right side is $\frac{\lambda\ell}{\varepsilon_0}$.

$$E2\pi r\ell = \frac{\lambda\ell}{\varepsilon_0}$$

$$E = \frac{\lambda}{2\pi\varepsilon_0 r}$$

You should not be too surprised to see that this is the same expression as for the long line of charge or for the single long cylinder of radius a. Because $\lambda = \frac{q}{L}$, this result could also be written as

$$E = \frac{q}{2\pi\varepsilon_0 rL}$$

The charge on the outer cylinder does not contribute to the field between the two charged cylinders.

c. To model $b < r << L$, the radius of the Gaussian surface would be enlarged to surround both cylinders. Now the charge inside the Gaussian cylinder is zero, so there is no field. The restriction $r << L$ is necessary to ensure that the field is everywhere perpendicular to the curved part of the Gaussian cylinder. Note that such a combination of equal and oppositely charged concentric cylinders is also useful as a capacitor.

2.6 Nonconductors

To this point, you have encountered many examples with the primary emphasis on conductors. An important feature of conductors is that excess static charge cannot be held in position within them. If you wish to have a static charge distributed within a solid body, you must use a nonconducting substance. Let's analyze such a problem situation.

Example 10 A nonconducting sphere of radius a contains a positive charge of uniform density ρ. Find:

a. the total charge Q contained within the radius a

b. the expression for the electric field for $r > a$

c. the expression for the electric field for $0 < r < a$

Solutions First, let's be familiar with useful notation. The symbol ρ represents a volume charge density—that is, a charge per unit volume. Its unit is C/m^3. (You are already familiar with the concepts of linear density λ, in C/m, and with surface density σ, in C/m^2.) For constant charge density,

$$\rho = \frac{q}{V}$$

in which q represents the charge contained in a volume V. For a volume less than that of the entire sphere, the charge contained is

$$q = \rho V_{partial}$$

For the entire sphere, the total charge is given by Q where

$$Q = \rho V_{total}$$

a. The volume of a sphere is $4/3\ \pi r^3$. For the whole sphere the volume is $4/3\ \pi a^3$, so, the total charge is

$$Q = \frac{4}{3}\pi \rho a^3$$

b. For $r > a$, you will need a spherical Gaussian surface of radius r, concentric with the charged sphere, as shown on the left of Figure 2.23. To calculate, start with Gauss's Law:

$$\oint_{\substack{closed\\surface}} \vec{E} \cdot d\vec{A} = \frac{q_{inside}}{\varepsilon_0}$$

$$E(4\pi r^2) = \frac{4\pi \rho a^3}{3\varepsilon_0}$$

$$E = \frac{\rho a^3}{3\varepsilon_0 r^2}$$

Note that E is once again as it would be for a point charge; it varies as $\frac{1}{r^2}$.

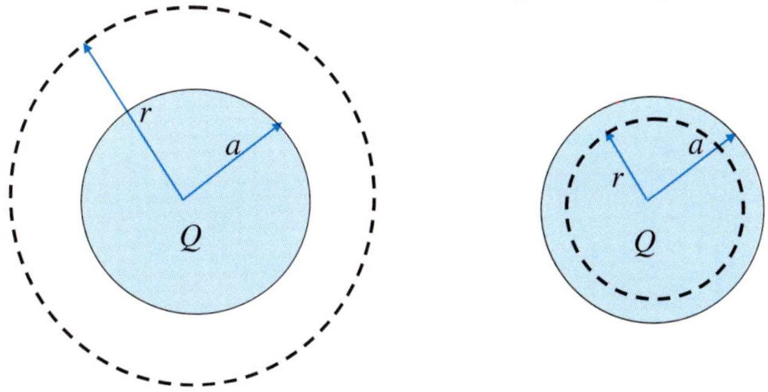

Figure 2.23

c. For $0 < r < a$, you need a new representation, as on the right of Figure 2.23. The left side of Gauss's equation remains $E(4\pi r^2)$. However, the charge inside the smaller Gaussian surface is no longer the total charge. Because ρ is uniform, the charge inside is the product of ρ and the smaller volume $\frac{4}{3}\pi r^3$

$$q_{inside} = \frac{4}{3}\pi r^3 \rho$$

and Gauss's Law becomes

$$E(4\pi r^2) = \frac{4\pi \rho r^3}{3\varepsilon_0}$$

$$E = \frac{\rho r}{3\varepsilon_0}$$

The electric field is now directly proportional to r. It grows as r increases because the amount of charge inside the radius r increases.

Here is an important technique to remember: whenever you derive a result in symbols, always check the units of your answer. In this case, the answer $E = \frac{\rho r}{3\varepsilon_0}$ has units $\frac{\text{C}}{\text{m}^3}\text{m}\frac{\text{N} \cdot \text{m}^2}{\text{C}^2}$, which reduce to N/C, the correct unit for electric field. By this practice, you can be reasonably assured that you did not drop an important constant at some step. Another good idea is to check whether the two expressions for the electric field—both inside and outside the charged sphere—give the same value at the surface of the sphere, when $r = a$. Here, both $E = \frac{\rho r}{3\varepsilon_0}$ and $E = \frac{\rho a^3}{3\varepsilon_0 r^2}$ do give $E = \frac{\rho a}{3\varepsilon_0}$ when $r = a$.

Example 11 A nonconducting sphere of radius a contains a charge with density given by $\rho = \alpha r$, where α is a positive constant. Find

a. the total charge Q contained within the radius a

b. the expression for the electric field for $r > a$

c. the expression for the electric field for $0 < r < a$

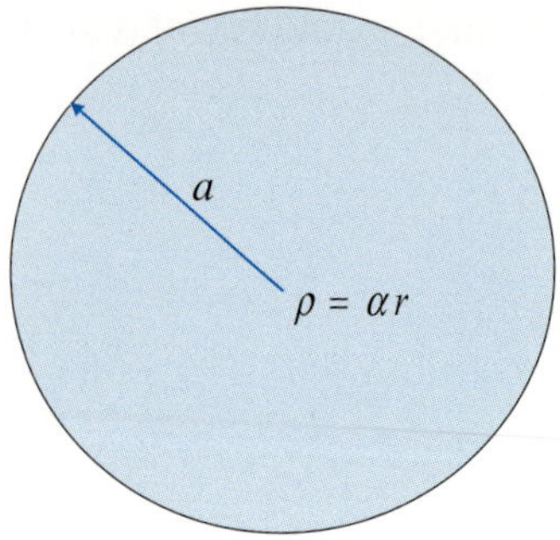

Figure 2.24

Solutions **a.** With a variable density, you will no longer be able to use $Q = \rho V_{total}$ to find the total charge. You must use the methods of calculus to sum up the different parts of the sphere with varying densities. Think of the sphere as consisting of concentric hollow shells, like the layers of an onion. Each shell has a volume dV equal to its own surface area times its thickness dr, as illustrated in Figure 2.25.

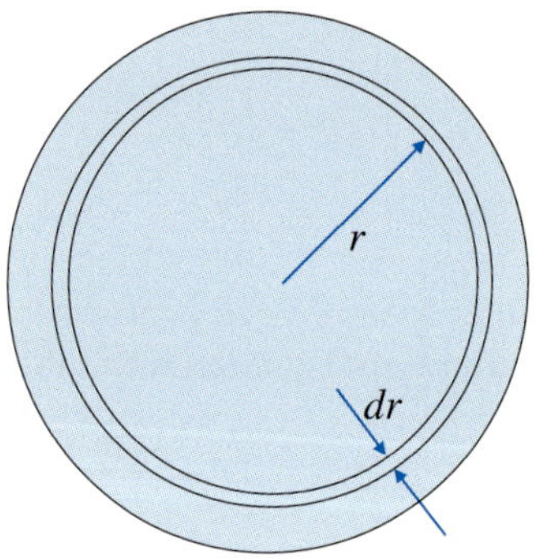

Figure 2.25

The charge on each shell is dq, where

$$dq = \rho dV$$

$$dq = \alpha r\left(4\pi r^2 dr\right)$$

$$dq = 4\pi\alpha r^3 dr$$

You can now find the total charge by integrating over r from 0 to a.

$$Q_{total} = \int_0^a 4\pi\alpha r^3 dr = 4\pi\alpha \int_0^a r^3 dr$$

$$Q_{total} = \left[\frac{4\pi\alpha r^4}{4}\right]_0^a$$

$$Q = \pi\alpha a^4$$

b. For $r > a$, the electric field behaves like that of a point charge Q. Substituting the total charge Q above for the sphere, we get

$$E = \frac{Q}{4\pi\varepsilon_0 r^2} = \frac{\alpha a^4}{4\varepsilon_0 r^2}$$

c. For $0 < r < a$, begin with Gauss's Law again.

$$\oint_{\substack{closed\\surface}} \vec{E} \cdot d\vec{A} = \frac{q_{inside}}{\varepsilon_0}$$

You can find q_{inside} by replacing a with r in the upper limit integral expression for Q_{total}. So you have

$$E\left(4\pi r^2\right) = \frac{\pi\alpha r^4}{\varepsilon_0}$$

$$E = \frac{\alpha r^2}{4\varepsilon_0}$$

To check, let $r = a$. The expressions for E in parts **b** and **c** are the same:

$$E = \frac{\alpha a^2}{4\varepsilon_0}$$

A graph of E vs. r looks like this:

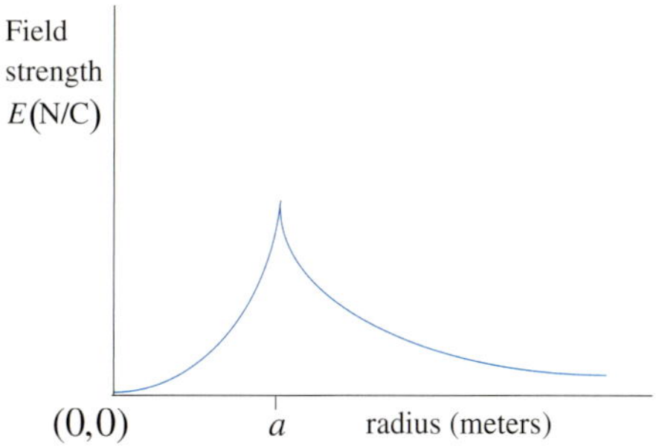

Figure 2.26

Example 12 A long nonconducting cylinder of radius R contains a charge of constant density ρ, as shown in Figure 2.27. Find

a. the charge contained in a length ℓ of the cylinder.

b. the magnitude of the electric field, where $r > R$, for a point close to the charged cylinder and not too near the end.

c. the magnitude of the electric field, where $r < R$, for a point not too near the end of the charged cylinder.

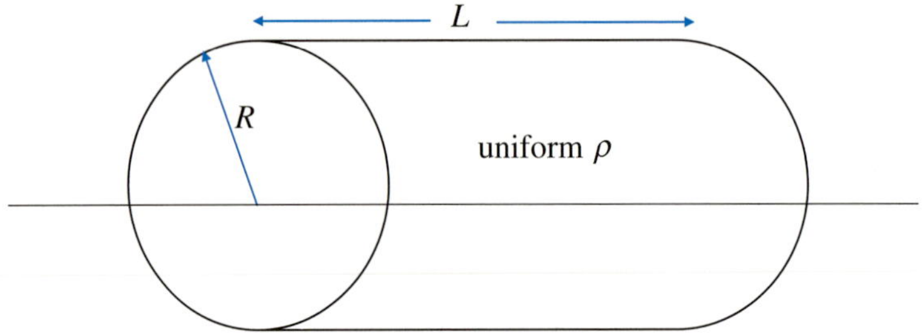

Figure 2.27

Solutions

a. The charge contained in a length ℓ is just the charge density ρ multiplied by the volume of the cylinder. The volume of the part of the cylinder of radius R and length ℓ is $\pi R^2\ell$. The charge in length ℓ is then

$$q = \pi\rho R^2\ell$$

b. Look at the Gaussian surface drawn in Figure 2.21. Write down Gauss's Law:

$$\oint_{\substack{\text{closed}\\ \text{surface}}} \vec{E} \cdot d\vec{A} = \frac{q_{inside}}{\varepsilon_0}$$

$$E(2\pi r\ell) = \frac{\pi\rho R^2\ell}{\varepsilon_0}$$

$$E = \frac{\rho R^2}{2\varepsilon_0 r}$$

c. Look at the Gaussian surface drawn in Figure 2.20. Use Gauss's Law:

$$\oint_{\substack{\text{closed}\\ \text{surface}}} \vec{E} \cdot d\vec{A} = \frac{q_{inside}}{\varepsilon_0}$$

The left-hand side produces the usual $E(2\pi r\ell)$. The charge inside the Gaussian surface is just the density times the volume enclosed by the Gaussian surface. Putting this together you get

$$E = 2\pi r\ell = \frac{\rho\pi r^2\ell}{\varepsilon_0}$$

$$E = \frac{\rho r}{2\varepsilon_0}$$

Checking the units of E, checking for agreement between the two expressions for the electric field at $r = R$, and constructing a graph of E vs. r, will be left as exercises for you to complete.

Exercises

1. **a.** Sketch the electric field pattern for the region around a point charge of magnitude $+3q$ located a short distance from a point charge of magnitude $-2q$.

   b. Draw any Gaussian surface enclosing both charges and verify that the net number of lines emerging from the surface is equal to half the number of lines entering the charge of $-2q$.

2. A long rod covered with a uniform linear charge density (charge/meter) of $+\lambda$ runs down the center of a long, uncharged cylindrical conducting shell, as depicted in Figure 2.28.

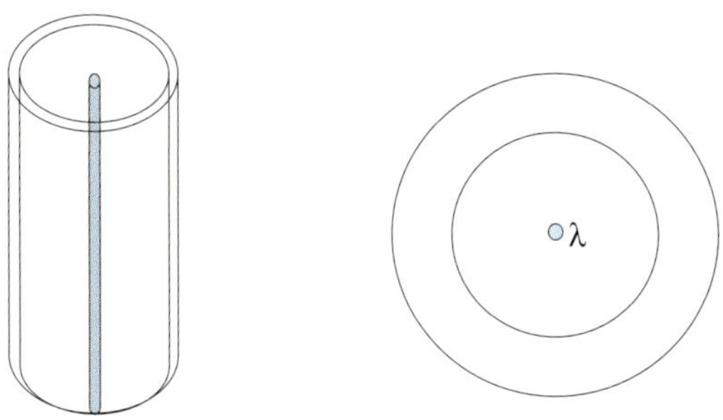

Figure 2.28

   a. Using the end-on view, sketch the electric field pattern both inside and outside the shell. Indicate the induced charge distributions on the inner and outer surfaces of the shell by means of plus and minus signs.

   b. The charged rod is now moved off center by $R/4$. Repeat part **2a**.

3. Verify that the units of the solutions for E in Example **12b** and **12c** are correct. Also check that when $r = R$, the two solutions are identical.

4. Graph E vs. r for Examples 10 and 12.

Problems

1. Using the methods of calculus, as they were employed in Example 2 of Chapter 1, find the magnitude of the field at a distance a from a long line of charge with linear charge density λ. Verify that you get the same result as in Example 7 of this chapter, $E = \frac{\lambda}{2\pi\varepsilon_0 a}$, where the same problem was solved using Gauss's Law. (*Hint*: see Figure 2.18 and imagine the line to consist of little charges of magnitude dq.)

2. In Example 12, the electric field intensity for a long solid nonconducting cylinder of radius R, with a uniform charge density ρ spread through its entire volume, was found to be $E = \frac{\rho r}{2\varepsilon_0}$ inside the cylinder and $E = \frac{\rho R^2}{2\varepsilon_0 r}$ outside the cylinder. A related problem is to find the expressions for the electric field due to a long nonconducting cylindrical shell of inner radius R_1 and outer radius R_2 with a uniform charge density ρ. Find E for

   a. $r < R_1$

   b. $R_1 < r < R_2$

   c. $r > R_2$

   See Figure 2.29 for a side view and an end view of the shell.

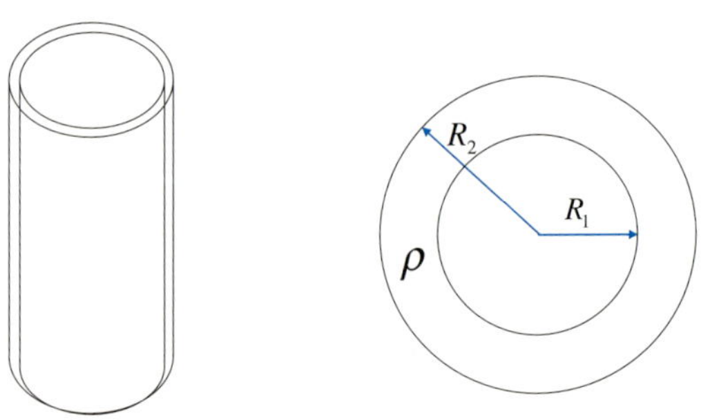

Figure 2.29

3. A nonconducting sphere and a long nonconducting cylinder are positioned as in Figure 2.30. Both have the same constant charge density through their volumes. The magnitude of the charge density is ρ and the charge is negative on both objects. The sphere has radius r_s and the cylinder radius r_c. The distance between their centers is r.

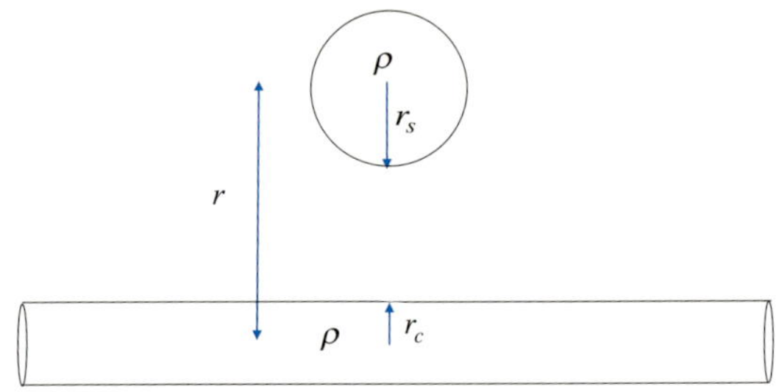

Figure 2.30

   a. Find the electric field at a point halfway between their centers that is not inside either the sphere or the cylinder. Which way will it point?

   b. In the case where $r_c = r_s = r/2$, so that the sphere and cylinder are almost touching, evaluate your answer to part **a.** Which way will it point?

Problems

4. Two identical spheres of radius R carry a negative charge Q spread uniformly over their volume. The spheres lie on the x-axis with their centers separated by $4R$, as shown in Figure 2.31. Calculate the x- and y-components of the electric field at the following points:

   a. $(0, 0)$

   b. $(2R, 0)$

   c. $(2R, 2R)$

   d. $\left(\frac{1}{2}R, 0\right)$

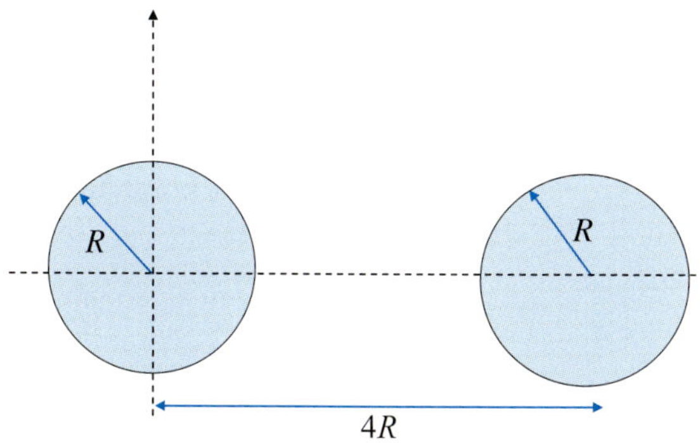

Figure 2.31

5. A long solid nonconducting cylinder has a charge distribution given by

$$\rho = \rho_0 \frac{R - r}{R}$$

where ρ_0 is a positive constant. The radius of the cylinder is R.

Find the magnitude of the electric field at a point a distance r from the axis of the cylinder where

   a. $r > R$

   b. $r < R$

   c. Describe the direction of the electric field in both of these regions.

6. A solid conducting sphere of radius a is surrounded by a concentric conducting shell of inner radius b and outer radius c, as shown in Figure 2.32. The sphere has a charge $+2Q$, and the shell has a charge $-Q$.

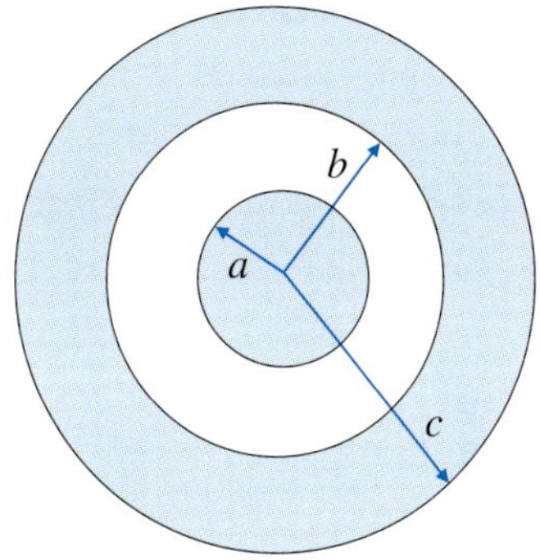

Figure 2.32

   a. Write an expression for the magnitude of the electric field for $a < r < b$.

   b. What is the value of the electric field for $b < r < c$?

   c. Write an expression for the magnitude of the electric field for $r > c$.

   d. How much charge, in terms of Q, is located on the inner and outer surfaces of the conducting shell?

   e. Sketch a graph of E vs. r for this problem.

Problems

7. An infinite sheet of nonconducting material has a uniform surface charge density $+\sigma$. An infinite conducting plate that has no net charge is placed parallel to the sheet. The charge on the conducting plate redistributes itself so that it looks as in Figure 2.33, with surface charge densities of $+\frac{\sigma}{2}$ and $-\frac{\sigma}{2}$ on the faces. Find the magnitude and direction of the electric field in regions I, II, and III in terms of σ and universal constants.

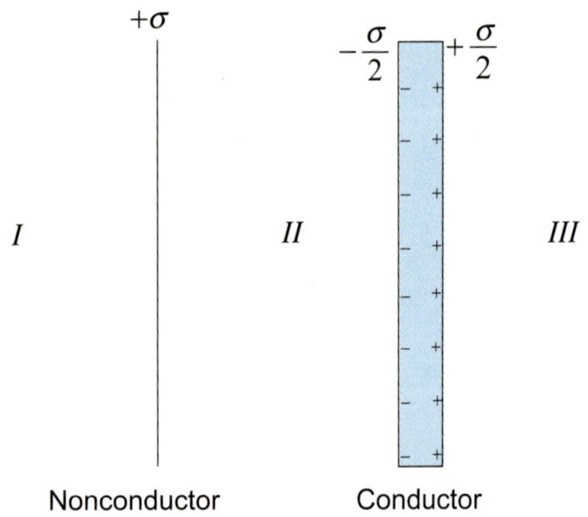

Figure 2.33

CHAPTER 3

Electric Potential and Potential Difference

3.1 Review of Work and Potential Energy

The first time you saw the term **potential energy** in Physics was probably in the context of gravitational potential energy. So let's start with a simple gravitational example that we can then extend to the world of charges. This example will put a few basic intuitive ideas in place before we launch into a more mathematical treatment of the topic. Imagine taking your physics book from your desk and, as illustrated in Figure 3.1, raising it slowly to a height Δh.

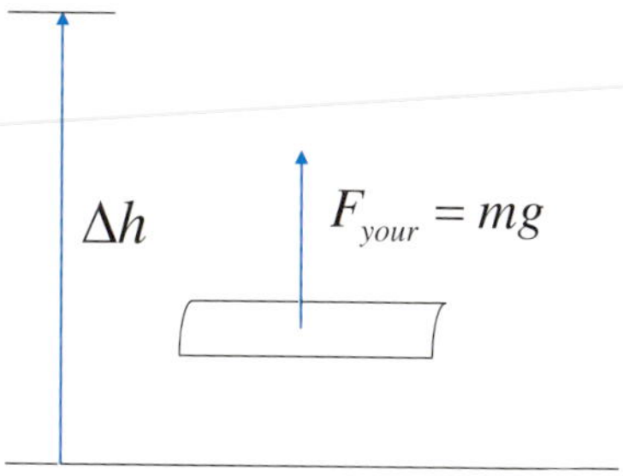

Figure 3.1

As you raise the book, you are doing work. For a constant force,

$$W = \vec{F} \cdot \Delta\vec{s}$$

where $\vec{F}$ is your force and $\Delta\vec{s}$ is a vector representing the displacement of the book. Your force is approximately constant and equal to the weight of the book. If you raise the book straight up, the magnitude of $\Delta\vec{s}$ is equal to Δh, the change in height. Also, $\vec{F}$ and $\Delta\vec{s}$ are parallel, so the cosine in the dot product is equal to 1. Making these substitutions, the work W becomes

$$W = mg\Delta h$$

The right side of this equation may now be understood as the change in the gravitational potential energy, ΔU. In words, the equation may be read, "The work you do is equal to the change in the gravitational potential energy."

Let's imagine another example, this one electrical, where work is understood in exactly the same way. To replace the constant force of gravity, you will need a constant electrical force. You can establish one by setting up a system of charged parallel plates with equal and opposite charges, as illustrated in Figure 1.9 of Chapter 1. That drawing can be rotated by 90° to appear as in Figure 3.2 (even though it will work just as well in any orientation).

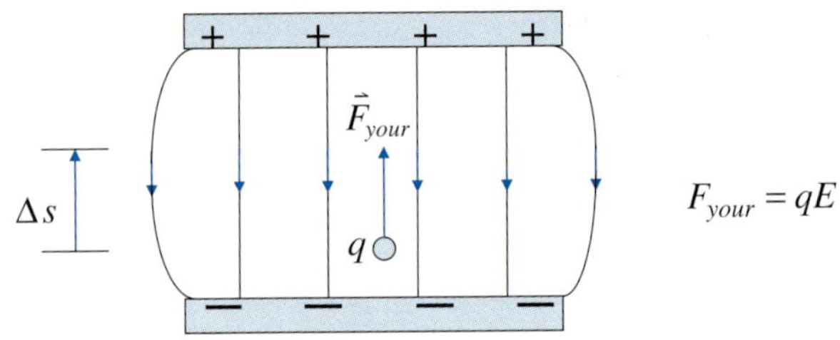

Figure 3.2

Now imagine taking a small positive charge q and pushing it upward in the field, through a distance $\Delta\vec{s}$. The force you must apply is the same amount of force the electric field would apply to this charge, namely qE. Your force is applied in the opposite direction to the force applied by the electric field, just as your force in lifting the book was in the opposite direction to the force of gravity. The work you do will then be the product of force and distance, which is

$$W = qE\Delta s$$

As in the gravitational case, the right side of the equation may be understood as the change in the **electrical potential energy**, so you may read the equation as, "The work that you do is equal to the change in the electrical potential energy."

A constant force is the simplest possible case in problems of potential energy. In a gravitational situation, you may have seen the additional complication of an object being moved a great distance in the field of Earth, so that the gravitational field $\vec{g}$ is no longer a constant. To make this case more concrete, picture your moving a space capsule a displacement $\Delta\vec{s}$ in the field of Earth, as shown in Figure 3.3.

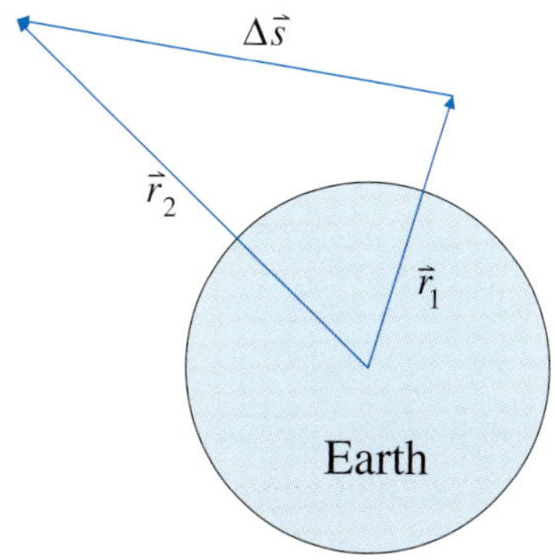

Figure 3.3

You will recall that it does not matter which path you follow to get from the position indicated by radius $\vec{r}_1$ to the position indicated by radius $\vec{r}_2$. You would have to do an equal amount of work regardless of whether you first went straight out along a radial line, and then around in a circular path to get to the new position shown. No work needs to be done on the circular path because the opposing force of gravity and the displacement are perpendicular on this path. All the work is done on the radial part of the trip. You can write the gravitational force in vector form as

$$\vec{F}_g = m_{capsule}\vec{g} = -\frac{Gm_{capsule}m_{Earth}}{r^2}\hat{r}$$

where the minus sign indicates that the force of gravity is directed in the opposite direction to that of $\hat{r}$. Gravity is directed inward, and the radial unit vector is directed outward. In this situation, however, you need to consider your force acting to push the capsule away from Earth. This force is in the outward direction and, therefore, is given by

$$\vec{F}_{your} = \frac{Gm_{capsule}m_{Earth}}{r^2}\hat{r}$$

Because the force is not constant, the work must now be calculated from the general formula for the work done by a force $\vec{F}$

$$W = \int_{path} \vec{F} \cdot d\vec{s}$$

This line integral says that the total work done by the force as an object moves along a certain path can be calculated by summing up the little bits of work $dW = \vec{F} \cdot d\vec{s}$ done over each little displacement $d\vec{s}$ along the path. As noted earlier, for the gravitational force, the work done between two endpoints does not depend on the path taken between the endpoints.

For this example, $d\vec{s} = dr\hat{r}$; that is, the path is along a radius beginning at $\vec{r}_1$ and ending at $\vec{r}_2$. Thus you can write

$$W = \int_{path} \vec{F}_{your} \cdot d\vec{s} = \int_{r_1}^{r_2} \vec{F}_{your} \cdot \hat{r}dr = \int_{r_1}^{r_2} F_{your}dr$$

$$W = Gm_{capsule} m_{Earth} \int_{r_1}^{r_2} \frac{dr}{r^2}$$

$$W = -Gm_{capsule} m_{Earth} \left[\frac{1}{r_2} - \frac{1}{r_1} \right]$$

Given then that the potential energy at r_2 is $-\frac{Gm_{capsule} m_{Earth}}{r_2}$, with a similar expression for r_1, the last equation may be understood as, "The work that you do becomes the change in the gravitational potential energy." Notice that this equation for the work does give a positive answer when $r_2 > r_1$.

The electrical case most closely analogous to that of the space capsule and Earth is the case of two point charges opposite in sign. The force that q_1 exerts on q_2 is given by Coulomb's Law

$$\vec{F}_e = \frac{1}{4\pi\varepsilon_0} \frac{q_1 q_2}{r^2} \hat{r}$$

where the unit vector $\hat{r}$ is directed outward from q_1 along the line connecting the two charges. This force will be attractive because the product $q_1 q_2$ is negative. The force that you have to supply to move q_2 slowly away from q_1 is just the negative of the Coulomb force:

$$\vec{F}_{your} = -\frac{1}{4\pi\varepsilon_0} \frac{q_1 q_2}{r^2} \hat{r}$$

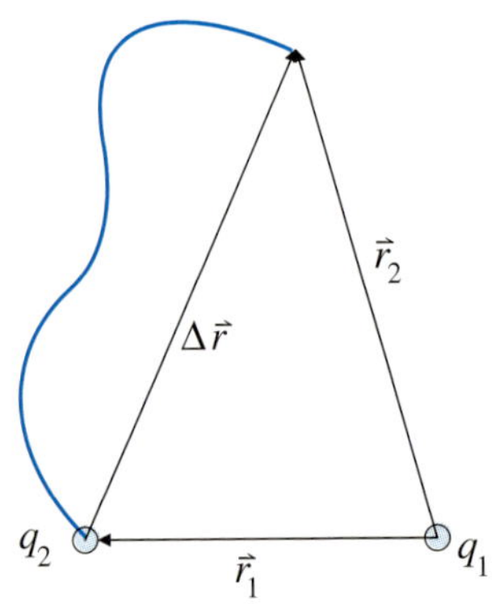

Figure 3.4

The charge q_2 is to be moved along the arbitrary curved path from a radius $\vec{r}_1$ to a radius $\vec{r}_2$, as illustrated in Figure 3.4. The work you would have to do is the same as the work needed to push the charge directly along a radial path from $\vec{r}_1$ to $\vec{r}_2$ with $d\vec{s} = dr\hat{r}$.

$$W = \int_{path} \vec{F}_{your} \cdot d\vec{s} = \int_{r_1}^{r_2} \left(-\frac{1}{4\pi\varepsilon_0} \frac{q_1 q_2}{r^2} \hat{r} \right) \cdot dr\hat{r}$$

$$W = -\frac{1}{4\pi\varepsilon_0} q_1 q_2 \int_{r_1}^{r_2} \frac{dr}{r^2}$$

$$W = \frac{1}{4\pi\varepsilon_0} q_1 q_2 \left[\left(\frac{1}{r_2}\right) - \left(\frac{1}{r_1}\right)\right]$$

When $r_2 > r_1$, the value in square brackets will have a negative value. However, because the product $q_1 q_2$ is also negative, the work is positive, as is the change in electrical potential energy. Because this equation shows a difference between similar looking terms, it is convenient to define the electrical potential energy for two point charges by the equation

$$U = \frac{q_1 q_2}{4\pi\varepsilon_0 r}$$

Whenever you use numerical values for the charges, enter them in with their signs and you will get the correct sign for the potential energy.

As a check on this scheme for potential energy, you may want to consider what happens when the charges have the same sign. How could you increase their potential energy? You could push them closer together, bringing q_2 inward from $\vec{r}_2$ to $\vec{r}_1$ along the line connecting the two charges, so that $d\vec{s} = dr\hat{r}$.

$$W = \int_{path} \vec{F}_{your} \cdot d\vec{s}$$

$$W = -\frac{1}{4\pi\varepsilon_0} q_1 q_2 \int_{r_2}^{r_1} \frac{dr}{r^2}$$

$$W = \frac{1}{4\pi\varepsilon_0} q_1 q_2 \left[\frac{1}{r_1} - \frac{1}{r_2}\right]$$

The value in square brackets would now have a positive sign because $r_2 > r_1$, but the product $q_1 q_2$ would also be positive. Once again, the whole value for the work you do is positive, as is the change in potential energy.

If you then release the charge q_2, the field will accelerate the charge. As it reaches greater and greater radial distances from the other charge, its kinetic energy will increase and its potential energy must decrease. To have a clear idea of what is going on in any given situation, remember this: the change in potential energy is the work you have to do to push the charge slowly from one point in an electric field to another. If the field is doing the work and not you, the potential energy decreases. If you have to do the work, the potential energy increases.

To analyze potential energy for electric field further, let's take another look at the equation

$$U = \frac{q_1 q_2}{4\pi\varepsilon_0 r}$$

You might want to know where the potential energy is zero. As r becomes very large, the value of U approaches zero asymptotically. Sometimes we say U is zero at ∞. What does U vs. r look like graphically? That depends on the signs of the charges, as seen in Figure 3.5. Remember that when $U < 0$, an increase in potential energy will bring its value closer to zero.

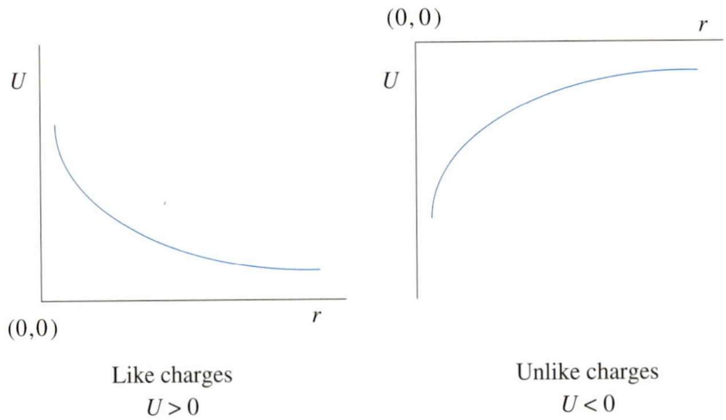

Figure 3.5

Example 1 Two point charges $q_1 = +3.0 \times 10^{-6}$ C and $q_1 = -4.0 \times 10^{-6}$ C are located at the coordinates (0,0) and (2,0), respectively, as illustrated in Figure 3.6.

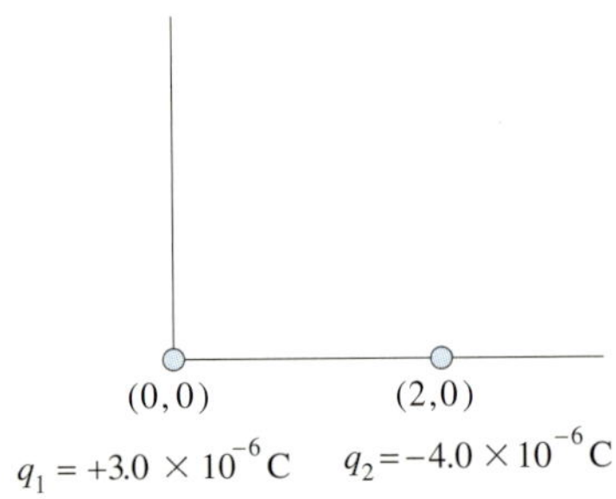

Figure 3.6

a. Find the potential energy of the system of two charges.

Solution

$$U = \frac{q_1 q_2}{4\pi\varepsilon_0 r}$$

$$U = (9.0 \times 10^9)(3.0 \times 10^{-6})(-4.0 \times 10^{-6})/2.0$$

$$U = -0.054\text{J}$$

The potential energy is negative and as a measure of energy is, of course, in Joules. The sign makes sense if you think of the problem as having started out with the charges very far apart, and then ask: How much work would you have to do to push them to these coordinates? Actually, because the charges are attracting each other, you would have to apply your force opposite to the direction of their displacements; that is, you would do negative work to hold them back. In this case, the field would do the positive work. The potential energy would be zero when the charges were very far apart, and now it is negative because of the negative work you do to slowly reposition them.

b. A third charge $q_3 = -2.0 \times 10^{-6}$ C is now located at (0,2), as shown in the next illustration. Find the potential energy of this system of charges.

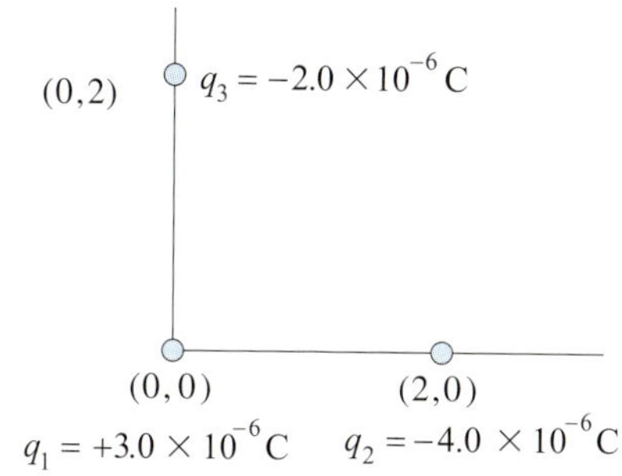

Figure 3.7

Solution

You will need to think of the third charge as being located initially very far away where it has no potential energy, then being moved in to the coordinates (0,2). The work needed to do this is just the sum of the work needed to move q_3 in the field of the first charge q_1 and the work needed to move q_3 in the field of the second charge q_2.

$$U = \frac{q_1 q_3}{4\pi\varepsilon_0 r_1} + \frac{q_2 q_3}{4\pi\varepsilon_0 r_2}$$

$$U = (9 \times 10^9)(-2.0 \times 10^{-6})\left(\frac{3.0 \times 10^{-6}}{2.0} + \frac{-4.0 \times 10^{-6}}{2.8}\right)$$

$$U = -1.3 \times 10^{-3}\ \text{J}$$

Once again, the negative sign indicates that you would have to do negative work to move the third charge slowly from infinity to the final position. The electric field does positive work. You will also notice that because potential energy is a scalar, you do not need to think about directions in such a problem. The numerical values may simply be added together. You do need to keep in mind that the equation

$$U = \frac{q_1 q_2}{4\pi\varepsilon_0 r}$$

is only to be used for point charges, and that the zero of potential energy is assumed to be at ∞. For other charge distributions, you will need the more general formulation

$$\Delta U = \int_{path} \vec{F}_{your} \cdot d\vec{s}$$

Let's make one more substitution to make this equation conform to the way it is usually presented in textbooks. The force $\vec{F}_{your}$ that you exert in slowly moving the charge is just the negative of the electric force $F_{electric}$ that all the other charges exert on the charge. Thus you can write

$$\Delta U = -\int_{path} \vec{F}_{electric} \cdot d\vec{s}$$

3.2 Potential and Potential Energy with Applications to Point Charges

You will recall from Section 1.2 of Chapter 1 that the magnitude of the force between two charges is written as

$$F = \frac{q_1 q_2}{4\pi\varepsilon_0 r^2}$$

In this equation, q_1 and q_2 represent the magnitudes of the charges. The magnitude of the electric field created by the charge q_1 is calculated by dividing the force exerted on q_2 by the magnitude of q_2.

$$E = \frac{F}{q_2} = \frac{q_1}{4\pi\varepsilon_0 r^2}$$

The direction of $\vec{E}$ is defined as the direction of the force on a small positive charge placed in the field. Because you are dividing out the value of q_2, you can think of the field as the force per unit charge. In Chapter 1 we went from the concrete idea of a force to the more abstract idea of a field permeating space. The advantage of the field concept becomes apparent in the many examples you have seen. If you have an array of charges on a plane, a sphere, a cylinder or any other shape, and you calculate the electric field produced by that array, then the force on any given charge q you bring into that field will be found from $\vec{F} = q\vec{E}$.

This technique is so useful that it is worth extending to the concept of electric potential energy as well. The electric potential energy of a point charge depends on the value of the charge. It would be useful to define a quantity that does not depend on the particular charge placed in the field but which can be used to calculate the potential energy of any charge placed in the field. To see how this can be done, let's start with the potential energy of two point charges

$$U = \frac{q_1 q_2}{4\pi\varepsilon_0 r}$$

then define a new quantity, the **electric potential** V due to q_1, by dividing this equation by q_2:

$$V = \frac{q_1}{4\pi\varepsilon_0 r}$$

Now an arbitrary charge q placed a distance r from q_1 will have potential energy $U = qV$.

The electric potential, like the potential energy, is zero at ∞. The **potential difference** between two points in the space around a point charge q will be

$$\Delta V = V_2 - V_1 = \frac{q}{4\pi\varepsilon_0}\left(\frac{1}{r_2} - \frac{1}{r_1}\right)$$

The difference in potential between two points does not depend on where the zero of potential is located. If you know the electric potential difference, then you can calculate the change in the potential energy for any charge q:

$$\Delta U = q\Delta V$$

Electric potential, like potential energy, is a scalar quantity. Its units are the Joule/Coulomb (J/C), commonly called the **volt (V)**. It is important to keep in mind that electric potential is the potential energy per unit charge associated with an electric field. You must still think of doing work to push a charge from infinity to the point where you want to evaluate the potential. The difference is that in calculating electric potential, you assume the charge being moved is 1 C—that is, a unit charge.

Let's look at part b of Example 1 again, but now from the point of view of electric potential.

Example 2 Calculate the potential due to q_1 and q_2 at the point P with coordinates (0,2).

Solution The situation is essentially the same as in Figure 3.7, except that the magnitude of the third charge is not specified. The potential at point P would be the sum of the potential due to charge q_1 and the potential due to charge q_2.

$$V = \frac{1}{4\pi\varepsilon_0}\left(\frac{q_1}{r_1} + \frac{q_2}{r_2}\right)$$

$$V = (9 \times 10^9)\left(\frac{3.0 \times 10^{-6}}{2.0} + \frac{-4.0 \times 10^{-6}}{2.8}\right)$$

$$V = 6.4 \times 10^2 \text{V}$$

If you then wanted to know the potential energy associated with bringing in a third charge $q_3 = -2.0 \times 10^{-6}$C from a point very far away from point P, you would simply multiply this potential V by the third charge

$$U = q_3 V$$

Your result would be the same as the answer to Example 1b.

Example 3 Find the potential at the center C of the square array of point charges in Figure 3.8.

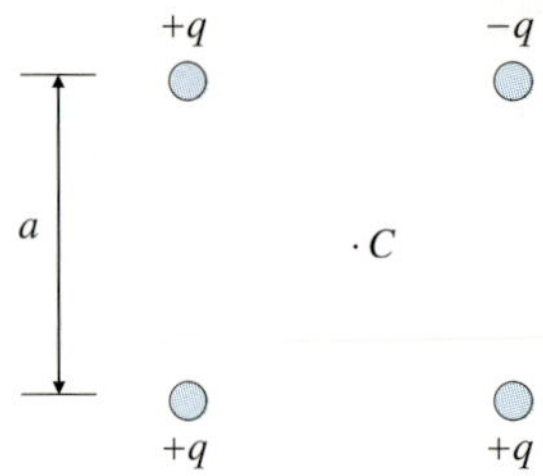

Figure 3.8

Solution You must write down the potential each individual charge would produce at point C and add them up. For each charge, the potential at C may be calculated from the point charge equation

$$V = \frac{q}{4\pi\varepsilon_0 r}$$

The distance r from each charge to point C is half the diagonal of the square, or $a\sqrt{2}/2$. For the four charges,

$$V = \frac{q}{4\pi\varepsilon_0 a\sqrt{2}/2}(+1+1+1-1) = \frac{q}{\sqrt{2}\pi\varepsilon_0 a}$$

Rationalizing the radical in the denominator gives

$$V = \frac{\sqrt{2}q}{2\pi\varepsilon_0 a}$$

Example 4 For the square array shown in Figure 3.9, where is the potential equal to zero?

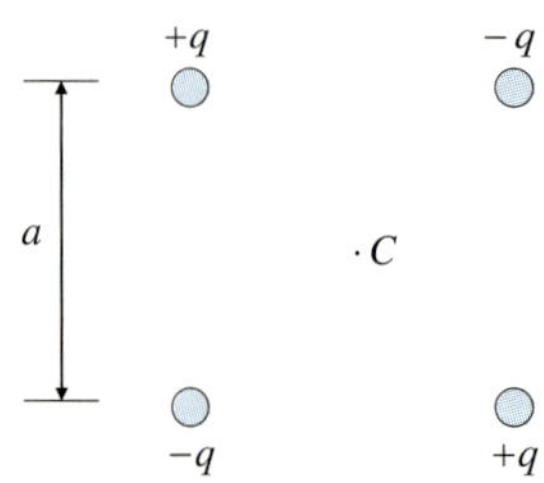

Figure 3.9

Solution The potential is certainly zero at point C, because there are now two positive and two negative point charges and they are all at the same distance from C. There are also other points where a pair of opposite charges are equidistant; for example, anywhere on a vertical line through the center or on a horizontal line through the center, as illustrated in Figure 3.10.

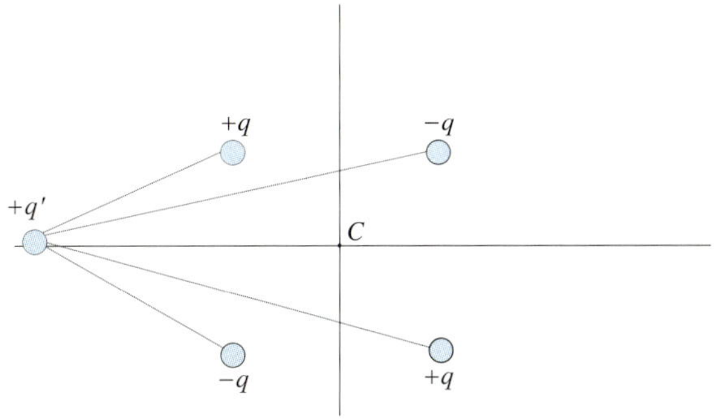

Figure 3.10

You see that a positive charge q' placed on the horizontal line would be equidistant from the positive and negative pair on the left, and also from the positive and negative pair on the right, anywhere along that line. The same would be true of any point on the vertical line through C.

But note: because the potential is zero along these lines does not mean that the electric field is zero everywhere along them. The four electric field vectors at the location of q' would look like the ones shown in Figure 3.11. The net field, indicated by the bold arrow, is in the negative y direction, perpendicular to the line. Because the dot product of perpendicular vectors is zero, no work can ever be done on a charge moving along this line. This does not mean that a charge could be given a little kick along this line and it would sail right through the center. Because the net electric field is in the negative y direction, the charge would be deflected in this direction.

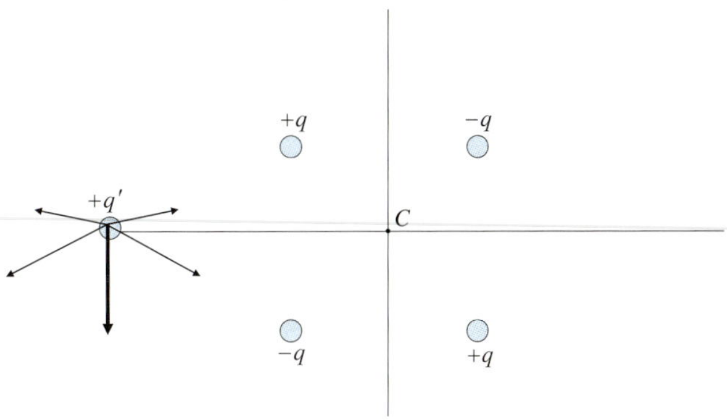

Figure 3.11

In general, you can write the electric potential difference between two points explicitly in terms of the electric field in the region. Start with the equation for the change in electric potential energy of a charge q.

$$\Delta U = -\int_{path} \vec{F}_{electric} \cdot d\vec{s}$$

Recall that $\vec{F}_{electric} = q\vec{E}$. If you simply divide both sides of this equation by q, you will get the general result

$$\Delta V = -\int_{path} \vec{E} \cdot d\vec{s}$$

3.3 Examples Involving Gauss's Law

Let's return a moment to the large nonconducting plate of charge first discussed in Example 2 of Chapter 1, and later in Example 4 of Chapter 2. In this situation, you would not want to define the potential to be zero at ∞. If you approached from ∞ pushing your charge against a constant electric field, the potential would be

$$\Delta V = -\int_{path} \vec{E} \cdot d\vec{s} = -E\Delta s = \infty$$

It makes more sense in this case to define the zero of potential to be at the surface of the plate (similarly to the gravitational case, where the zero of potential energy for small changes in height is defined as the floor or the table top). Then the potential at a point a distance Δs from the plate is simply

$$\Delta V = V - 0 = -\int_{path} \vec{E} \cdot d\vec{s} = -E\Delta s$$

If the plate is positive, all potentials will be negative in sign because you do not have to push a positive charge to move it away from the plate. If the plate is negative, you do have to push a positive charge to move it away from the plate, so the sign of the potential is positive.

What happens when you have a pair of oppositely charged plates, as in Example 6 of Chapter 2? Usually, the negative plate is designated as the zero of potential and the potential rises linearly as you go across the gap to the positive plate. Figure 3.12 shows this scheme for a practical situation where a 3.0-volt battery is connected between the two plates.

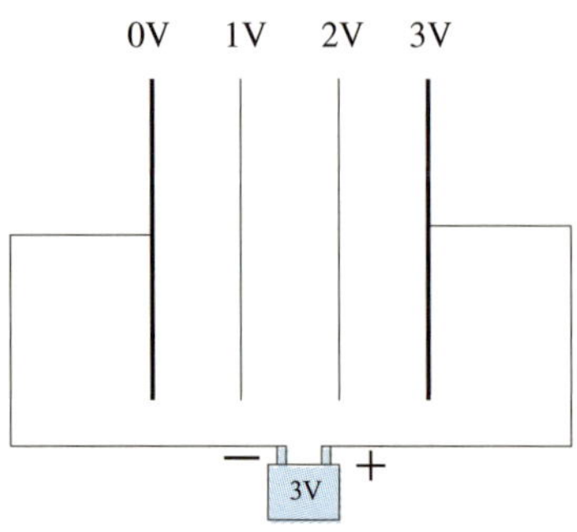

Figure 3.12

The lines labeled 1V and 2V represent large plane imaginary surfaces parallel to the plates themselves, for which the potential has a constant value. These are called **equipotential surfaces**. Equipotential surfaces are always perpendicular to the field at every point in space. Hence, you do not have to do any work to move a charge on such a surface. If this is starting to sound familiar, it is because the surfaces of conductors discussed earlier are examples of equipotential surfaces.

Example 5 A hollow conducting sphere of radius a has a charge $+Q$. Assume the potential is zero at ∞.

a. Find the potential at a point where $r > a$.

b. Find the potential at a point where $r < a$.

c. Graph the potential and the electric field as functions of r.

d. What are the shapes of the equipotential surfaces outside the sphere?

Solutions

a. Outside the sphere, the field is like that of a point charge. As usual, choose an integration path that follows a radius so that $d\vec{s} = dr\hat{r}$.

$$\vec{E} = \frac{1}{4\pi\varepsilon_0}\frac{Q}{r^2}\hat{r}$$

$$\Delta V = -\int_{path} \vec{E}\cdot d\vec{s}$$

$$\Delta V = -\frac{Q}{4\pi\varepsilon_0}\int_{\infty}^{r}\frac{\hat{r}\cdot dr\hat{r}}{r^2}$$

$$V - 0 = -\frac{Q}{4\pi\varepsilon_0}\left(-\frac{1}{r} + 0\right)$$

$$V = \frac{Q}{4\pi\varepsilon_0 r}$$

This expression is also the same as the potential at a distance r from a point charge Q.

b. For $r < a$, the field is zero, but that does not mean the potential is zero. You would still have to push the small positive charge all the way from ∞ to the radius a. The potential at the surface of the sphere will be

$$V = \frac{Q}{4\pi\varepsilon_0 a}$$

Then, if you imagine a small hole opening up in the sphere, to let the charge in, no more work must be done to get the charge all the way to $r = 0$. So, the potential anywhere inside the sphere is given by the equation

$$V = \frac{Q}{4\pi\varepsilon_0 a}$$

c. Graphs of V vs. r and E vs. r would appear as follows:

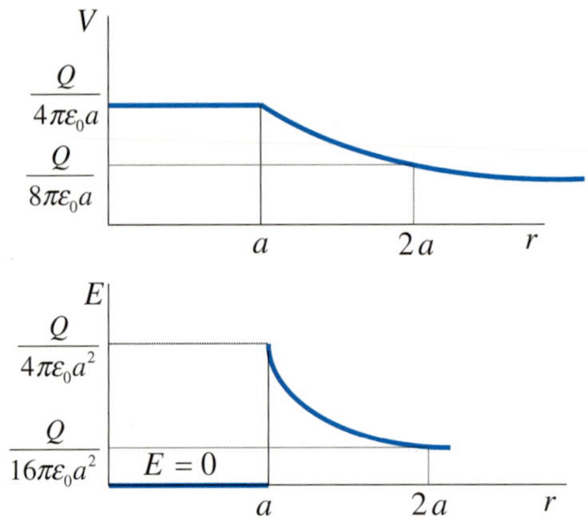

Figure 3.13

d. The equipotential surfaces are spherical because V depends on r and not on any angle.

Example 6 A solid nonconducting sphere of radius a has a uniform positive charge density ρ throughout its volume. The sphere is surrounded by a concentric conducting shell of inner radius b and outer radius c that carries no excess charge. The situation is as shown in Figure 2.8 of Chapter 2.

a. Find the total charge on the inner sphere.

b. Find the charge, including the sign, on the inner and outer surfaces of the shell.

c. Find the electric field for $r > c$, $b < r < c$, $a < r < b$ and $r < a$.

d. Find the potential at $r = c$, $r = b$, $r = a$ and the expression for the potential when $r < a$.

e. Sketch the graphs of both field and potential as functions of r.

Solutions a. The total charge on the inner sphere will be the volume charge density multiplied by the volume:

$$Q = \rho\left(\frac{4}{3}\pi a^3\right) = \frac{4\pi\rho a^3}{3}$$

b. The inner surface of the conducting shell will have the same amount of charge, but it will be negative. This is necessary to satisfy Gauss's Law and keep the field inside the conductor zero. Because the conductor is neutral, to conserve charge, it must have $+Q$ on its outer surface.

c. Outside a spherical charge distribution, the field will be the same as for a point charge with the same total charge. The electric fields are as follows:

- For $r > c$, $E = \frac{1}{4\pi\varepsilon_0}\frac{4\pi\rho a^3}{3r^2} = \frac{\rho a^3}{3\varepsilon_0 r^2}$
- For $b < r < c$ (inside a conductor), $E = 0$
- For $a < r < b$, $E = \frac{\rho a^3}{3\varepsilon_0 r^2}$ (the same mathematical form as for $r > c$)
- For $r < a$, the charge enclosed by a Gaussian surface of radius r will be the total charge times the fraction of the volume enclosed by the surface:

$$q = \left(\frac{4\pi\rho a^3}{3r^2}\right)\frac{r^3}{a^3} = \frac{4\pi\rho r^3}{3}$$

You can use Gauss's Law then to find the field inside.

$$E4\pi r^2 = \frac{1}{\varepsilon_0}\left(\frac{4\pi\rho r^3}{3}\right)$$

$$E = \frac{\rho r}{3\varepsilon_0}$$

d. To find the potentials you will want to use

$$\Delta V = -\int_{path} \vec{E}\cdot d\vec{s}$$

Choose integration paths that follow a radius centered on the charge distribution, so that $\vec{ds} = dr\hat{r}$:

- For $r = c$, think of coming in from infinity to a radius c working against a field $E = \frac{\rho a^3}{3\varepsilon_0 r^2}$.

$$\Delta V = V_c - 0 = -\int_{\infty}^{c}\frac{\rho a^3}{3\varepsilon_0 r^2}dr$$

$$V_c = -\frac{\rho a^3}{3\varepsilon_0}\left[-\frac{1}{r}\right]_{\infty}^{c}$$

$$V_c = \frac{\rho a^3}{3\varepsilon_0 c}$$

- For the next part of the trip, to $r = b$, you are going through the conductor where there is no field. There can be no further change in potential. When you arrive at radius b, the potential is the same: $V_b = V_c$.
- For $r = a$, you must traverse the gap between the shell and the inner sphere. The field has the same form as on the outside but the limits are different.

$$\Delta V = V_a - V_b = -\int_{b}^{a}\frac{\rho a^3}{3\varepsilon_0 r^2}dr$$

$$V_a - \frac{\rho a^3}{3\varepsilon_0 c} = \frac{\rho a^3}{3\varepsilon_0}\left(\frac{1}{a} - \frac{1}{b}\right)$$

$$V_a = \frac{\rho a^3}{3\varepsilon_0}\left(\frac{1}{a} - \frac{1}{b} + \frac{1}{c}\right)$$

- For the last part of the trip, use the field for $r < a$, integrating from a to r.

$$\Delta V = V_r - V_a = -\int_a^r \frac{\rho r dr}{3\varepsilon_0}$$

$$V_r = \frac{\rho a^3}{3\varepsilon_0}\left(\frac{1}{a} - \frac{1}{b} + \frac{1}{c}\right) + \frac{\rho}{6\varepsilon_0}\left(a^2 - r^2\right)$$

You will notice that both terms are positive. This result must be so because you are finding the work per unit charge to push a positive charge against an electric field for the whole trip. At $r = 0$, the result becomes

$$V_0 = \frac{\rho a^3}{3\varepsilon_0}\left(\frac{1}{a} - \frac{1}{b} + \frac{1}{c}\right) + \frac{\rho}{6\varepsilon_0}\left(a^2\right) = \frac{\rho a^3}{3\varepsilon_0}\left(\frac{3}{2a} - \frac{1}{b} + \frac{1}{c}\right)$$

e. The graphs will look like those given in Figure 3.14.

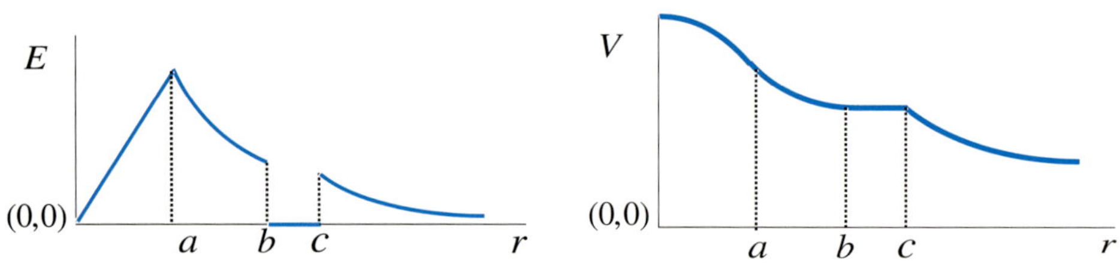

Figure 3.14

Example 7 Going back to Example 9 of Chapter 2, find the potential difference between the two cylinders there. The electric field between the cylinders for that example was found to be

$$E = \frac{\lambda}{2\pi\varepsilon_0 r}$$

Solution The potential difference will be

$$\Delta V = -\int_{path} \vec{E} \cdot d\vec{s}$$

$$\Delta V = -\int_b^a \frac{\lambda dr}{2\pi\varepsilon_0 r}$$

$$\Delta V = -\frac{\lambda}{2\pi\varepsilon_0}\left[\ln r\right]_b^a$$

$$\Delta V = \frac{\lambda}{2\pi\varepsilon_0} \ln\left(\frac{b}{a}\right)$$

You will note that this result is a positive number, as the charge on the inner cylinder is positive and $b > a$. The inner cylinder has the higher potential.

3.4 Summary of Important Results

Let's conclude with a brief summary of the important points that have been defined or derived in this book.

1. When a point charge is alone in empty space, it is surrounded by an electric field. The field at any given point in space is the force on a positive charge of magnitude 1 C placed at that point.
2. If multiple point charges are the source of the field, the resultant field may be calculated from the vector superposition of the fields of the individual point charges.
3. For continuous arrays of charge, such as lines, sheets or rings, it is possible to calculate the electric field using the methods of integral calculus. An integral is, after all, just a sum, although you must take into account that it is a vector sum.
4. The electric flux through a surface is defined as

$$\Phi = \oint_{surface} E \cdot d\vec{A}$$

5. Gauss's Law states that the electric flux through a closed surface is independent of the shape of the surface and depends only on the net charge inside the surface. The equation is

$$\oint_{\substack{closed\\surface}} \vec{E} \cdot dA = \frac{q_{inside}}{\varepsilon_0}$$

6. Gauss's Law has important consequences for conductors. There cannot be any electric fields inside conductors with static charge; the charge must reside on the outer surface of the conductors. Also, the field at the surface of a conductor is perpendicular to the surface.
7. Gauss's Law provides an efficient and simple method of calculating the electric field for certain symmetrical arrays of charge, particularly those with spherical, cylindrical or planar symmetry. Some care must be exercised when dealing with conductors, because the presence of another charge may result in a rearrangement of the charge. The charge distribution must be known before Gauss's Law can be applied.
8. The change in electric potential energy is the work that you do in moving a charge from one point to another in an electric field.

9. The electric potential at a point in space is the work that you do to move a positive charge of 1 C from a point where the potential is defined as zero to the specified point. For point charges and spheres, the zero of potential is usually taken to be at infinity. For long cylinders or large sheets of charge, it is usually taken to be on one of the charged surfaces.

Exercises

1. Assuming the potential is zero at infinity for the square array of charges in Figure 3.15, where else is the potential zero?

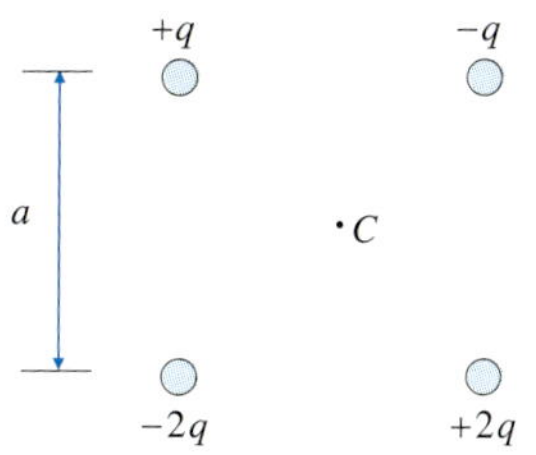

Figure 3.15

2. Find an expression for the electric potential

   a. at the center of a thin ring of charge of radius a and constant positive charge density λ, as in Figure 3.16a.

   b. at the center of an arc of charge with a central angle of 90°, as in Figure 3.16b.

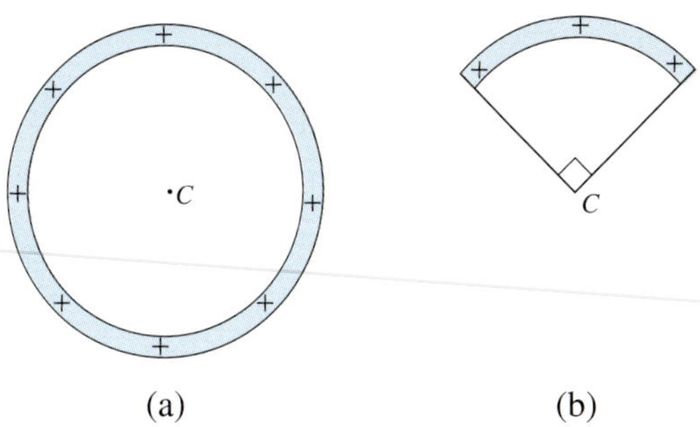

Figure 3.16

3. For the charges given in Figure 3.17:

   a. in which region(s) of the x-axis is it possible to have a point with zero electric field?

   b. in which region(s) of the x-axis is it possible to have $V = 0$ for some finite x, assuming the zero of potential at infinity?

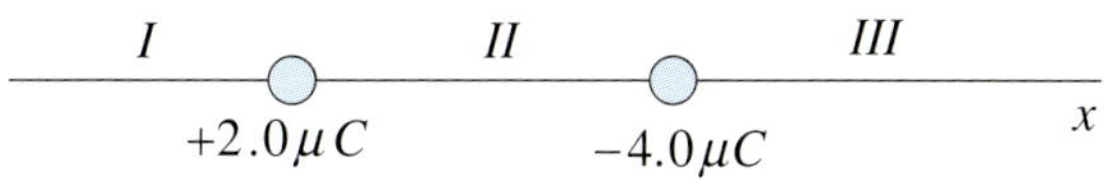

Figure 3.17

4. A large hollow conducting sphere is charged so that it has a potential of 500 V relative to infinity.

   a. What is the electric field inside the sphere?

   b. What is the electric potential inside the sphere?

5. If the electric field is zero at a particular point in space, does it follow that the electric potential must also be zero at that point? Give an example to justify your answer.

Exercises

6. Two parallel conducting plates have equal and opposite charges and are positioned as in Figure 3.18.

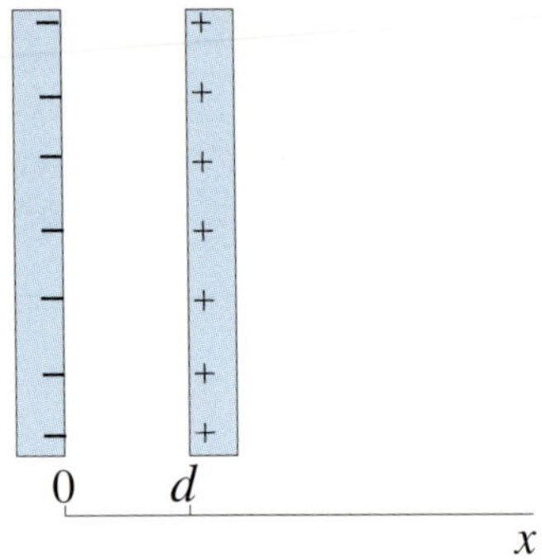

Figure 3.18

Sketch graphs of the electric field and the electric potential as functions of x. For the potential graph, assume $V = 0$ at the negative plate.

7. A negative point charge moves toward a large flat plate, also charged, in empty space. The negative charge speeds up as it gets closer to the charged plate.
   a. What is the sign of the charge on the plate?
   b. Does the potential energy of the system increase or decrease? Explain.
   c. Does the electric potential due to the plate increase or decrease as you get closer to it? Explain.

8. Sketch the equipotential curves for the electric field diagram in Figure 3.19.

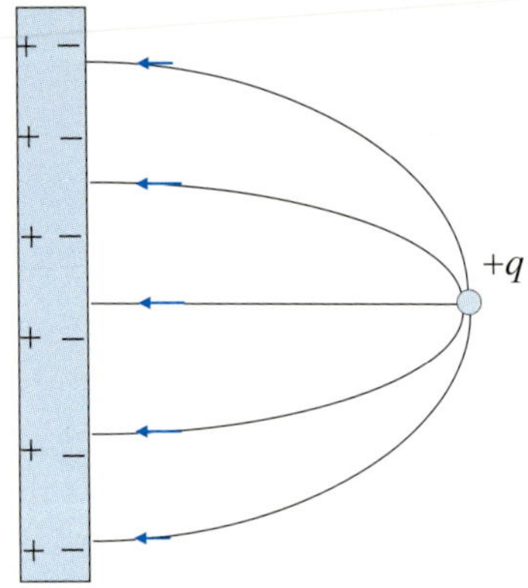

Figure 3.19

Problems

1. A charge $+Q$ is located at (0,0) and a charge $-2Q$ is located on the x-axis at (2,0).

   a. Locate the point(s) on the x-axis where $V = 0$ relative to infinity.

   b. Locate the point(s) on the y-axis where $V = 0$ relative to infinity.

2. Find an expression for the potential V

   a. at point P on the y-axis for the system of point charges in Figure 3.20. Assume $V = 0$ at infinity. Verify that this answer is reasonable by evaluating it at $y = 0$ and at $y >> a$.

   b. at point R on the x-axis. Verify that the answer seems reasonable for $x >> a$.

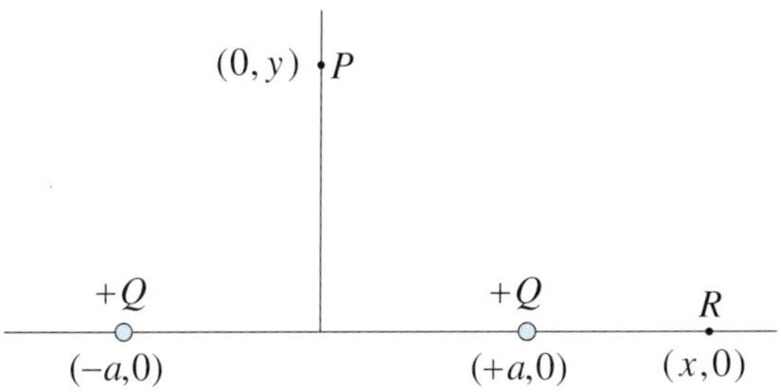

Figure 3.20

3. Two point charges are located as shown in Figure 3.21.

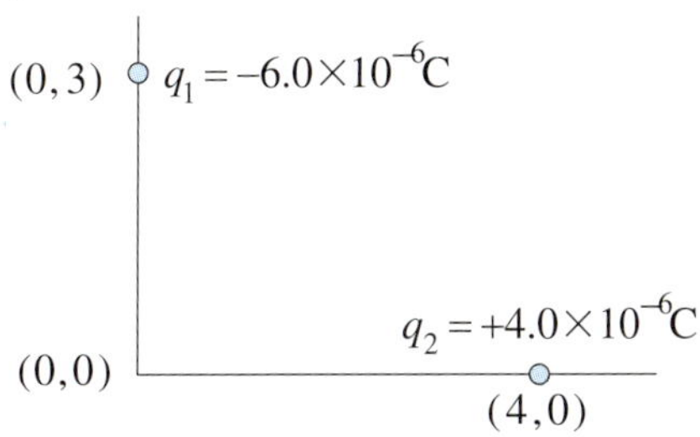

Figure 3.21

   a. Find the potential at the origin. Assume $V = 0$ at infinity.

   b. How much work would be required to bring a charge $q_3 = -2.0 \times 10^{-6}$ C from infinity and place it at the origin?

4. A solid nonconducting sphere of radius a has a charge density ρ given by

$$\rho = \rho_0\left(\frac{a - r}{a}\right)$$

   a. Find the total charge Q contained by the sphere.

   b. Find the potential at the surface of the sphere relative to infinity.

   c. Find the potential difference between an interior point at radius r and the surface of the sphere.

   d. Find the potential of that point inside the sphere relative to infinity.

5. A nonconducting sphere of radius $a = 0.18\ m$ is at a potential of -1000 V relative to infinity.

   a. How much charge is on the sphere?

   The nonconducting sphere is then enclosed by an uncharged spherical conducting shell of inner radius $b = 0.20$ m and outer radius $c = 0.25$ m. The situation is like that given in Figure 2.8 of Chapter 2.

   b. How much charge will be induced on the inner and outer surfaces of the conducting shell?

   c. Find the new potential of a point on the inner sphere.

   d. Show that if the shell is very thin, the new potential approaches -1000V.

Solutions to Exercises and Problems*

Chapter 1 Exercises

1.

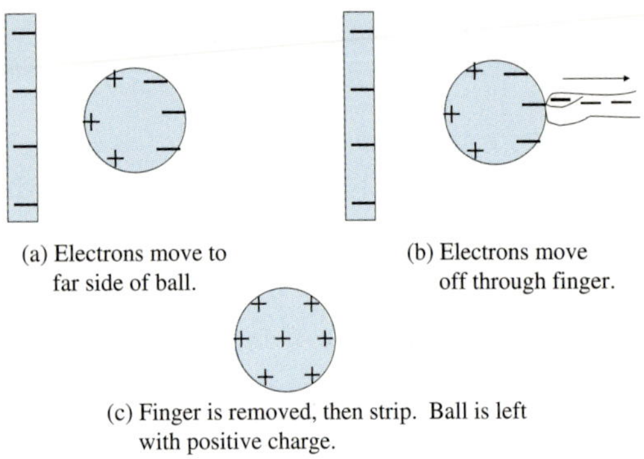

Figure S.1

3. $$\Phi = \oint \vec{E} \cdot d\vec{A}$$

For $\vec{E}$ parallel to $\vec{A}$, $\vec{E} \cdot \vec{A} = E(\pi R^2)$. From Example 2 of Chapter 1, recall that E has a constant value: $E = \sigma/2\varepsilon_0$. So,

$$\Phi = \frac{\sigma}{2\varepsilon_0}\,\pi R^2$$

For a 45° angle, $\Phi = \frac{\sqrt{2}}{2}\frac{\sigma}{2\varepsilon_0}\pi R^2 = \frac{\sqrt{2}\sigma\pi R^2}{4\varepsilon_0}$

For $\vec{E} \perp \vec{A}$, the field arrows do not enter or leave the loop: $\Phi = 0$

5. Symmetry tells us that because the charge is at the center of the cube, the flux must be the same for each face. A cube has 6 faces. Therefore,

$$\Phi = \frac{2.3 \times 10^5\ \text{N} \cdot \text{m}^2/\text{C}}{6} = 3.8 \times 10^4\ \text{N} \cdot \text{m}^2/\text{C}$$

7. If $E = 0$ everywhere on a closed surface, then

$$\oint \vec{E} \cdot d\vec{A} = 0$$

There is no flux through that surface. Therefore, Gauss's Law tells us that there is no net charge inside the surface.

Chapter 1 Problems

1. Figure S.3 shows the vectors at all three points.

a. At point P_1 both field vectors point in the positive x direction.

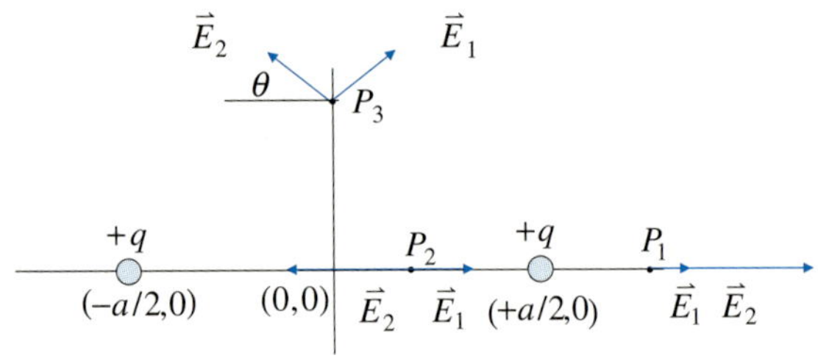

Figure S.3

$$\vec{E} = \frac{q}{4\pi\varepsilon_0}\left[\frac{1}{(x - a/2)^2} + \frac{1}{(x + a/2)^2}\right]\hat{\imath}$$

$$\vec{E} = \frac{q}{4\pi\varepsilon_0}\left[\frac{x^2 + ax + a^2/4 + x^2 - ax + a^2/4}{(x - a/2)^2(x + a/2)^2}\right]\hat{\imath}$$

$$\vec{E} = \frac{2q}{4\pi\varepsilon_0}\left[\frac{x^2 + a^2/4}{(x^2 - a^2/4)^2}\right]\hat{\imath}$$

When $x >> a/2$, but not infinitely great, you can neglect $\frac{a^2}{4}$ in the brackets. This gives

$$\vec{E} = \frac{2q}{4\pi\varepsilon_0 x^2}\,\hat{\imath}$$

which is the field of a point charge of magnitude $2q$. This is reasonable; from far away, the two charges behave as if they are right next to each other.

b. At point P_2 the field vectors are oppositely directed.

$$\vec{E} = \frac{q}{4\pi\varepsilon_0}\left[\frac{1}{(a/2 - x)^2} - \frac{1}{(a/2 + x)^2}\right]\hat{\imath}$$

$$\vec{E} = \frac{2q}{4\pi\varepsilon_0}\left[\frac{ax}{(a^2/4 - x^2)^2}\right]\hat{\imath}$$

When $x = 0$, the field is zero, which is reasonable, because the fields of the separate point charges are equal and opposite at this point.

* Solutions to all exercises and problems may be found on the publisher's website at http://www.peoplescollegeprep.com/gauss_law.

c. At point P_3 on the y-axis, the field vectors have equal and opposite x-components. Their y-components are equal to the product of the field due to each charge and $\sin\theta$. The y-components are also equal to each other, so the final result is twice one y-component.

$$\vec{E} = 2\left[\frac{q}{4\pi\varepsilon_0(a^2/4 + y^2)}\right](\sin\theta)\,\hat{j}$$

$$\vec{E} = 2\left[\frac{q}{4\pi\varepsilon_0(a^2/4 + y^2)}\right]\frac{y}{(a^2/4 + y^2)^{1/2}}\,\hat{j}$$

$$\vec{E} = \frac{2q}{4\pi\varepsilon_0}\,\frac{y}{(a^2/4 + y^2)^{3/2}}\,\hat{j}$$

When, $y >> a/2$, this reduces to

$$\vec{E} = \frac{1}{4\pi\varepsilon_0}\frac{2q}{y^2}\,\hat{j}$$

which is the field of a point charge of magnitude $2q$. This is just what you would expect, for the same reason given in the answer to **1a**.

3. For each little element of charge, $dq = \lambda dx$. The field at the origin due to this element is

$$dE = \frac{1}{4\pi\varepsilon_0}\frac{\lambda dx}{x^2} \text{ and } E = \frac{\lambda}{4\pi\varepsilon_0}\int_a^{a+L}\frac{dx}{x^2}$$

$$E = \frac{\lambda}{4\pi\varepsilon_0}\left(\frac{-1}{x}\right)_a^{a+L} = \frac{\lambda}{4\pi\varepsilon_0}\left(\frac{-1}{a+L} + \frac{1}{a}\right)$$

$$E = \frac{\lambda L}{4\pi\varepsilon_0 a(a + L)}$$

The field points away from the line of charge, in the negative x direction. Note that if $a >> L$, this reduces to the field of a point charge of magnitude λL.

Chapter 2 Exercises

1. a.

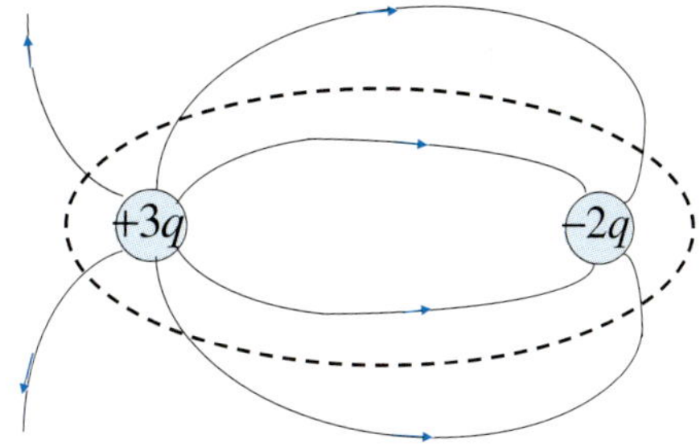

Figure S.5

b. In Figure S.5, the diagram of this particular Gaussian surface, 4 lines leave the Gaussian surface and 2 enter it, for a net number of lines leaving of 2, which is half the number of lines entering the negative charge (4). Note that six lines represent $3q$, four lines represent $2q$. Two lines leaving represent a charge of $+q$. This is the same as the net charge inside the Gaussian surface.

3. Check on correct units for solution to Example 12b: the units are

$$\left(\frac{C}{m^3}\right)\left(\frac{m^2}{m}\right)\left(\frac{Nm^2}{C^2}\right) = \frac{N}{C},$$

which is the correct unit for an electric field. Check on units for solution to Example 12c: the units are

$$\left(\frac{C}{m^3}\right)m\left(\frac{N\cdot m^2}{C^2}\right) = N/C,$$

which is correct for an electric field.

When $r = R$, the solution to Example 12b becomes

$$E = \frac{\rho R^2}{2\varepsilon_0 r} = \frac{\rho R^2}{2\varepsilon_0 R} = \frac{\rho R}{2\varepsilon_0}$$

which is the same as the solution to Example 12c at this radius.

Chapter 2 Problems

1. You will need to add some symbols to the diagram given in the original problem (Figure 2.18), so the diagram is given here again with these symbols added. As shown in Figure S.8, you will only need to consider the y-components of the $\vec{E}$ vectors at point P.

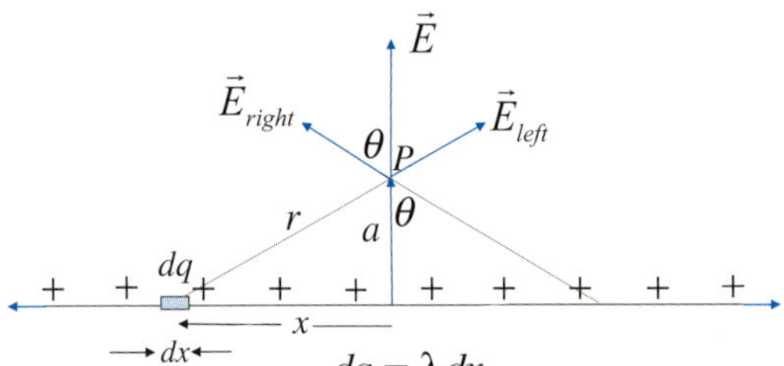

$$dq = \lambda dx$$

Horizontal contributions from symmetric bits of charge on either side of P will cancel.

Figure S.8

$$dE_y = \frac{1}{4\pi\varepsilon_0}\frac{dq}{r^2}\cos\theta$$

$$dE_y = \frac{1}{4\pi\varepsilon_0}\frac{\lambda dx}{r^2}\cos\theta$$

Change x and r into expressions involving θ.

$$x = a\tan\theta$$

$$dx = a\sec^2\theta d\theta$$

$$r^2 = a^2 + x^2 = a^2(1 + \tan^2\theta) = a^2\sec^2\theta$$

Substituting the expressions for dx and r^2 into the equation for dE_y gives

$$dE_y = \frac{1}{4\pi\varepsilon_0}\frac{\lambda a\sec^2\theta\cos\theta d\theta}{a^2\sec^2\theta}$$

$$dE_y = \frac{\lambda\cos\theta d\theta}{4\pi\varepsilon_0 a}$$

When dx is very far away, θ approaches $\pi/2$. Integrating from $-\pi/2$ to $+\pi/2$ gives

$$E = \frac{\lambda}{4\pi\varepsilon_0 a}\int_{-\pi/2}^{\pi/2}\cos\theta d\theta$$

$$E = \frac{\lambda}{4\pi\varepsilon_0 a}[\sin\theta]_{-\pi/2}^{+\pi/2}$$

$$E = \frac{\lambda}{4\pi\varepsilon_0 a}[1-(-1)]$$

$$E = \frac{\lambda}{2\pi\varepsilon_0 a}$$

This is the same answer as in Example 7 of Chapter 2, and clearly Gauss's Law provides a simpler route to the answer.

3 a. For the sphere, $E_s\left(4\pi\left(\frac{r}{2}\right)^2\right) = \frac{\rho\frac{4}{3}\pi r_s^3}{\varepsilon_0}$

$$E_s = \frac{4\rho r_s^3}{3\varepsilon_0 r^2}$$

For the cylinder, $E_c\left(2\pi\frac{r}{2}\ell\right) = \rho\left(\pi r_c^2\ell\right)/\varepsilon_0$

$$E_c = \frac{\rho r_c^2}{\varepsilon_0 r}$$

The net field is the difference between the two.

$$E = \frac{\rho}{\varepsilon_0}\left(\frac{4r_s^3}{3r^2} - \frac{r_c^2}{r}\right)$$

Each field points toward its own object because the charges are negative. The direction of the net field will be the direction of whichever field is greater, which in turn depends on the specific values of r_c and r_s.

b. If $r_c = r_s = r/2$, the field will be

$$E = \frac{\rho}{\varepsilon_0}\left(\frac{4r_s^3}{3(2r_s)^2} - \frac{r_s^2}{(2r_s)}\right)$$

$$E = \frac{\rho r_s}{\varepsilon_0}\left(\frac{1}{3} - \frac{1}{2}\right) = -\frac{\rho r_s}{6\varepsilon_0}$$

pointing toward the cylinder.

5. In order to use Gauss's Law for a nonuniform charge distribution, you must first find the charge inside the Gaussian surface. Choose a cylindrical surface coaxial with the cylinder. You can think of the cylinder as composed of many thin cylindrical shells, one of which is illustrated in Figure S.11.

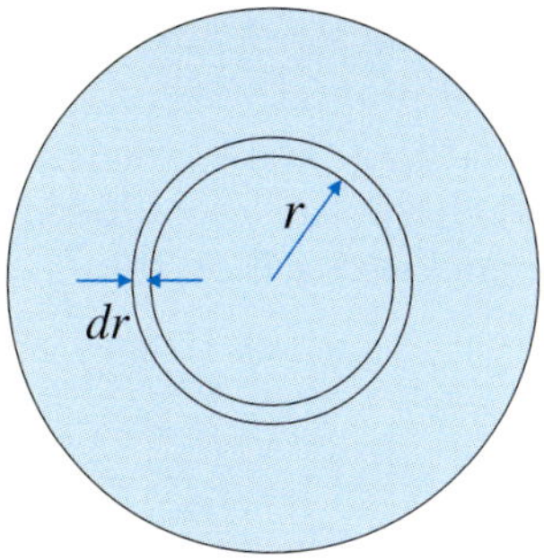

Figure S.11

The charge inside one of these shells is $dq = \rho dV$ where dV is the volume of the spherical shell.

$$V = \pi r^2 \ell$$

$$dV = 2\pi r \ell dr$$

Then the total charge inside a cylinder of radius r is

$$Q_{inside} = \int_0^r \rho dV$$

a. For $r > R$, you integrate all the way out to R to include all the charge.

$$Q_{inside} = \int_0^R \rho(2\pi r \ell dr)$$

$$Q_{inside} = \int_0^R \rho_0 \left(\frac{R-r}{R}\right)(2\pi r \ell dr)$$

$$Q_{inside} = \frac{2\pi\rho_0\ell}{R}\left[\int_0^R Rrdr - \int_0^R r^2 dr\right]$$

$$Q_{inside} = \frac{2\pi\rho_0\ell}{R}\left[\frac{Rr^2}{2} - \frac{r^3}{3}\right]_0^R$$

$$Q_{inside} = \frac{2\pi\rho_0\ell}{R}\left[\frac{R^3}{2} - \frac{R^3}{3}\right]_0 = \frac{\pi\rho_0\ell R^2}{3}$$

Applying Gauss's Law,

$$\oint_{\substack{closed\\surface}} \vec{E} \cdot d\vec{A} = \frac{q_{inside}}{\varepsilon_0}$$

$$E(2\pi r\ell) = \frac{\pi\rho_0\ell R^2}{3\varepsilon_0}$$

$$E = \frac{\rho_0 R^2}{6\varepsilon_0 r}$$

b. For $r < R$, the integral for the charge inside is the same, except that the upper limit is now r instead of R.

$$Q_{inside} = \frac{2\pi\rho_0\ell}{R}\left[\frac{Rr^2}{2} - \frac{r^3}{3}\right]_0^r = 2\pi\rho_0\ell\left[\frac{r^2}{2} - \frac{r^3}{3R}\right]$$

Applying Gauss's Law, as in part **a**,

$$E(2\pi r\ell) = \frac{2\pi\rho_0\ell}{\varepsilon_0}\left[\frac{r^2}{2} - \frac{r^3}{3R}\right]$$

$$E = \frac{\rho_0}{\varepsilon_0}\left[\frac{r}{2} - \frac{r^2}{3R}\right]$$

c. The electric field vector is perpendicular to the axis of the cylinder and directed away from the axis.

7. To find the field in region II, look at the Gaussian cylinder in Figure S.13.

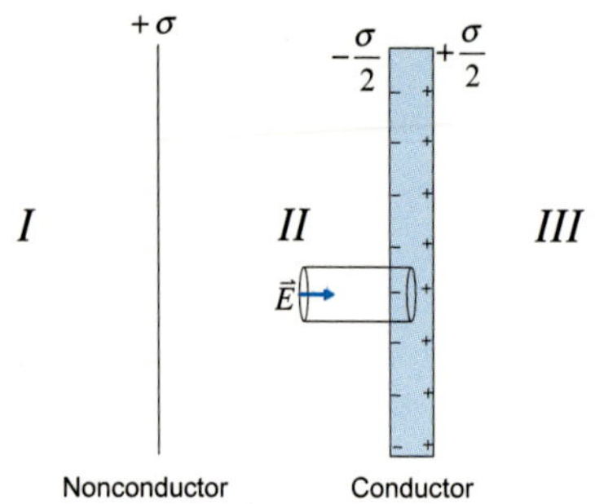

Figure S.13

There is no field inside the conductor, so the only field that contributes to the flux is the field through the base or cap of the cylinder in the region between the plate and the sheet. The charge enclosed is $\sigma A/2$, where A is the area of the end cap.

$$\oint_{\substack{closed\\surface}} \vec{E}\cdot d\vec{A} = \frac{q_{inside}}{\varepsilon_0}$$

$$EA = \sigma A/2\varepsilon_0$$

$$E = \frac{\sigma}{2\varepsilon_0}$$

The direction is toward the conducting plate, because fields always point from positive to negative charges.

Gaussian surfaces for regions I and III are shown in Figure S.14.

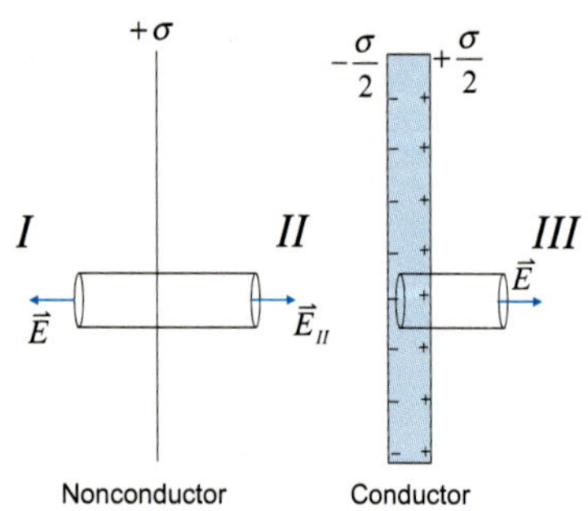

Figure S.14

For region III, again there is no field inside the conductor. The charge enclosed is again $\sigma A/2$. The calculation is the same as above.

$$E = \frac{\sigma}{2\varepsilon_0}$$

The vector points away from the conducting plate.

Finally, for region I, the illustration shows two contributions to the product of field and area. The flux is out of the cylinder on both ends, so these must be added together. The charge enclosed is now σA.

$$EA + \frac{\sigma A}{2\varepsilon_0} = \frac{\sigma A}{\varepsilon_0}$$

$$E = \frac{\sigma}{2\varepsilon_0}$$

pointing away from the nonconducting plate.

Chapter 3 Exercises

1. The potential is zero on a vertical line through the center of the square. Along this line the contributions to the potential from $+q$ and $-q$ will cancel, as will the contributions from $+2q$ and $-2q$.

3 a. As shown by the arrows in Figure S.15, it is not possible for the two fields to cancel out in region II because they are in the same direction. Neither is it possible for the two fields to cancel out in region III because any point will always be closer to the greater charge. This leaves only region I for a possible zero field point on the x-axis.

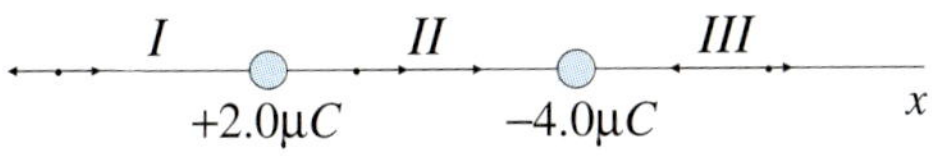

Figure S.15

b. Potential is not a vector, so to have zero potential at a point you only need the point to be twice as far away from the $-4.0\mu C$ charge as from the $+2.0\mu C$ charge. This could happen in either region I or region II.

5. If $E = 0$, it does not follow that V must be zero. One example is given in the preceding exercise. Another example is that of a system of two positive point charges Q a distance a apart. At the midpoint of the line joining the charges, the field is zero. The potential at this point is

$$V = \frac{1}{4\pi\varepsilon_0}\left(\frac{2Q}{a/2}\right) = \frac{Q}{\pi\varepsilon_0 a}$$

a positive number.

7 a. The negative charge speeds up, so it must be attracted to the plate. Opposite charges attract; therefore, the charge on the plate is positive.

b. The kinetic energy is increasing, so the potential energy must be decreasing. No work is being done on the charge other than that done by the field.

c. When talking about electric potential, you must assume a positive charge is being pushed toward the plate. You must do work to have this occur. Therefore, potential increases as you approach the positive plate. A negative charge will lose potential energy as it increases in electric potential.

Chapter 3 Problems

1 a. $V = V_1 + V_2 = 0$. In Exercise 3b you saw that this could happen at two points on the x-axis. See points A and B in Figure S.18.

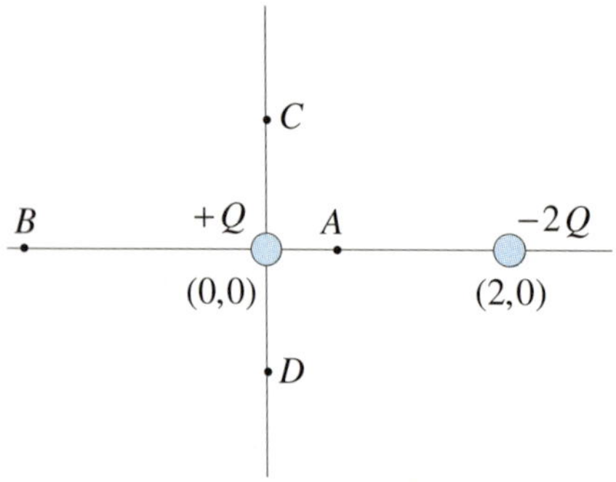

Figure S.18

For point A, each V is proportional to its charge over the distance of that charge from point A.

$$\frac{Q}{x} - \frac{2Q}{2-x} = 0$$

$$\frac{1}{x} = \frac{2}{2-x}$$

$$2x = 2 - x$$

$$x = 2/3$$

For point B, x is a negative number. The distances of each charge from point B must be positive numbers. The equation

$$V = V_1 + V_2 = 0$$

becomes

$$\frac{Q}{-x} - \frac{2Q}{2-x} = 0$$

$$-\frac{1}{x} = \frac{2}{2-x}$$

$$-2x = 2 - x$$

$$x = -2$$

A shortcut to these answers is possible if you realize that points A and B must be twice as far from the negative charge as from the positive one.

b. On the y-axis, symmetry tells us that there must be one point above the x-axis and another point just as far below. The distance from the negative charge to either point C or point D must be twice as great as the distance from the positive charge.

$$\sqrt{2^2 + y^2} = 2|y|$$

$$4 + y^2 = 4y^2$$

$$y = \pm\sqrt{\frac{4}{3}} = \pm\frac{2\sqrt{3}}{3}$$

3. a.

$$V = V_1 + V_2$$

$$V = \frac{1}{4\pi\varepsilon_0}\left(\frac{q_1}{r_1} + \frac{q_2}{r_2}\right)$$

$$V = (9.0 \times 10^9)\left(\frac{-6.0 \times 10^{-6}}{3.0} + \frac{4.0 \times 10^{-6}}{4.0}\right)$$

$$= -9.0 \times 10^3 \text{ V}$$

b. The work required to bring a charge in from infinity is the same as the potential energy of the system of charges after the work has been done. Once the potential due to the first two charges is known, you need only multiply that potential by the third charge to get the potential energy.

$$W = U = q_3V$$

$$W = (-2.0 \times 10^{-6} \text{ C})(-9.0 \times 10^3 \text{ V})$$

$$= 1.8 \times 10^{-2} \text{ J}$$

5. **a.** Using the point charge equation for potential relative to infinity

$$V = \frac{1}{4\pi\varepsilon_0}\frac{Q}{r}$$

and using the numbers given in the problem, we get

$$-1000 = \frac{9.0 \times 10^9\, Q}{0.18}$$

$$Q = -2.0 \times 10^{-8}\ \text{C}$$

b. As you have seen before, the charge on the shell becomes separated in such a way as to keep the field inside the conductor zero. The result is that the charge on the inner surface of the shell is $+2.0 \times 10^{-8}$ C and the charge on the outer surface of the shell is -2.0×10^{-8} C.

c. The field inside and outside the shell is given by the point charge equation

$$\vec{E} = \frac{1}{4\pi\varepsilon_0}\frac{Q}{r^2}\hat{r}$$

$$\vec{E} = 9 \times 10^9(-2.0 \times 10^{-8})\frac{\hat{r}}{r^2} = -1.8 \times 10^2\,\frac{\hat{r}}{r^2}$$

The potential on the surface of the sphere is found from integrating in the regions where the field is not zero.

$$V_a = -\int_{\infty}^{c} \vec{E}\cdot d\vec{r} - \int_{b}^{a} \vec{E}\cdot d\vec{r}$$

$$V_a = 1.8 \times 10^2\left[\int_{\infty}^{c}\frac{dr}{r^2} + \int_{b}^{a}\frac{dr}{r^2}\right]$$

$$V_a = 1.8 \times 10^2\left[-\frac{1}{c} - \frac{1}{a} + \frac{1}{b}\right]$$

$$= 1.8 \times 10^2(-4.0 - 5.6 + 5.0)$$

$$= -8.2 \times 10^2\ \text{V}$$

d. If the shell is very thin, b is approximately equal to c, and the potential reverts to

$$V = -1.8 \times 10^2\,\frac{1}{a} = -1000\ \text{V}$$

the same value it had before the shell was introduced.

Notes

Notes

Notes